THE
GROWTH
MINDSET
IN PROJECT MANAGEMENT

MAGDA JAWOROWICZ   PETER JAWOROWICZ

THE
GROWTH
MINDSET

in Project Management

Building a Culture of Learning, Agility, & Resilience in Every Project

San Francisco, 2025

Printed Worldwide
First Edition 2025
First Printing 2025

Paperback ISBN [I]: 979-8-9922630-0-8
Paperback ISBN [II]: 979-8-9922630-3-9
Hardcover ISBN [I]: 979-8-9922630-2-2
Hardcover ISBN [II]: 979-8-9922630-4-6
eBook ISBN: 979-8-9922630-1-5

Cover Design by Marlena Guderska
www.marlenaguderska.com

For Nathaniel, with love and a desire
to inspire his growth mindset.

Table of Contents

"I keep on making what I can't do yet
in order to learn to be able to do it."

Vincent Van Gogh
From a letter to Anthon van Rappard, August 1885

Introduction

Today's business environment is undergoing many transformations that are changing how companies operate, influencing strategic decisions, and fundamentally altering the work of project managers. With the acceleration of action, the increase in organizational power, and the rise of AI-powered collaboration, the future of project management is unfolding quicker than ever. Meanwhile, factors such as hybrid work environments, the use of AI tools, and new talent requirements radically change the way teams operate, leaders act, and projects get delivered. Both established enterprises and newer startups are becoming increasingly vulnerable to external shocks in today's environment, but not all are prepared to adapt swiftly.

Research conducted by McKinsey & Company shows that resilient companies have a solid competitive advantage, producing 50% more total shareholder value during recovery periods.[1] The reason for this is that these firms are growth mindset organizations: they prioritize empowering their employees, encouraging continuous learning, and creating an adaptable culture. In their effort to navigate these changes, companies must also keep pace with the demands to attract and retain talent, address capacity inefficiencies, and drive revenue, all while dealing with ever-increasing complexity. Nevertheless, at the time of writing this book, many organizations are not yet ready for this evolving environment, with 39% of workers considering quitting their jobs due to unrealistic expectations. Furthermore, most enterprises are not yet harnessing the full potential of the sustained culture transformation because they are not equipped to integrate it into their processes.

Encountering solutions for these emerging challenges is not just a matter of technical expertise—what is needed is the application of a growth mindset. This mindset allows project managers and teams to treat failure as a learning opportunity, fosters continuous improvement, and cultivates a company culture of learning, adaptation, and resilience. By adopting this attitude, leaders can break barriers, drive innovation, and nurture the fortitude needed to meet the challenges of the ever-changing business conditions. With uncertainty and complexity becoming the new normal for project managers in today's hyper-competitive landscape, traditional approaches must be transformed. Agility, resilience, and mindful leadership are now required traits in this dawning era of project management.

Staying up to speed with this interconnected environment is imperative for any business—from startups to Fortune 500 organizations—that wishes to master project management in order to meet their objectives, optimize processes, and deliver results that benefit stakeholders. Meanwhile, the data shows that the project management field is booming. According to the U.S. Bureau of Labor Statistics, in 2023, there were 973,600 project managers employed in the U.S. market, and it is forecasted that jobs in this area will grow 7% annually, which is an above-average growth rate.[2] This robust demand shows the increasing recognition of project management in driving organizational growth and adapting to challenges in the business context. Globally, there are 90 million workers employed in the project management area, and it is expected that this number will increase to 102 million by 2030.[3]

The impact of the rapidly changing business landscape on project management is multilayered. Unfortunately, there is no one-size-fits-all model. According to the Project Management Institute's *Pulse of the Profession* report, project managers need to adapt their skills to the diverse environments, problems, and strategic objectives of each project.[4] From industry features to organizational culture, project managers have never faced such high-pressure situations. Although project managers may have perfected their organization's methodologies, this alone is no longer sufficient to ensure successful project execution. Companies need to step out of the box and look for new ways to achieve improved project results. The Project Management Institute's forecast states the issue clearly: "To increase project performance rates, modern project management requires a combination of skills and competencies, along with the right culture, mindset and work environment".[5]

Magda and Peter Jaworowicz's book *The Growth Mindset in Project Management* looks at the life-changing effects of adopting a growth mindset. This inspiring work illustrates how the proper mindset and continuous learning can unlock innovation, unleash potential, and redefine success in project management today, helping leaders to forge the future with purpose and certainty. *The Growth Mindset in Project Management* explores the transformative effects of a growth mindset in managing projects. By moving from fixed-mindedness—in which skills are considered static—to growth-oriented thinking, project managers can instill resilience and creativity in their teams. The authors offer practical tips on how to build a learning-oriented, problem-solving, and flexible work environment that will enable your teams to navigate challenges effectively.

Moreover, this book underscores the need to tie the project to larger business priorities, leveraging data and performance indicators to make informed decisions and define success. It recommends the use of Agile and adaptive project management methodologies due to the fact that the ability to pivot and adapt is essential in the fast-paced business world. It explores how classical techniques, such as the Waterfall model, can be enhanced by a growth mindset, as well as the way hybrid models that combine the strengths of predictive and adaptive approaches can be more flexible and efficient. By stressing the importance of stakeholder management, the book provides tips for fostering strong collaboration between internal teams and external partners to ensure that everyone is on the same page.

Further, the book discusses continuous improvement methods—such as the Lean, Scrum, Six Sigma, and Kaizen approaches—and how they can be applied to automate tasks, eliminate waste, and improve project performance. It also discusses the application of innovation in the execution of projects, from MVPs and experimentation to using customer and user experience research to bring the project closer to market needs and expectations. By promoting these ideas, the book motivates project managers to learn continuously and to treat every problem as an opportunity for growth, both personally and collectively.

***

The first chapter, "Understanding the Growth Mindset," explores this approach, explains its roots, and discusses the ways it can be used in the context of project management. A growth mindset is embedded in the belief that abilities can be attained through commitment and effort. It differs from a fixed mindset, in which one's abilities are believed to be fixed or static. The shift to a growth mindset enables project managers to take risks and accept lessons learned. This mentality encourages flexibility, resilience, and a commitment to continuous learning—qualities that are essential for delivering outstanding outcomes.

Chapter 2, "Golden Rules of the Growth Mindset in Project Management," describes the fundamentals of growth-focused project management. The concept of "Golden Rules" highlights the importance of continuous improvement, learning from feedback, creating a collaborative, innovative culture, and thriving in the face of ambiguity. The chapter offers substantial guidance to help project managers apply these values to their daily work, emphasizing the importance of open communication and embracing challenges as opportunities. Examples of real-world cases illustrate how embracing these Golden Rules improves the performance of both projects and teams.

Chapter 3, "The Growth Mindset Ecosystem," delves into the dynamics of a growth mindset ecosystem, as well as how to cultivate a culture that nurtures it at the individual, project team, and organizational levels.

This chapter contrasts "Cultures of Genius," in which individuality is prized over collaboration, with "Cultures of Growth," in which cooperation and flexibility are rewarded. By developing a growth mindset across the organization, project teams can overcome obstacles with greater ease and employ creativity in their problem-solving.

Chapter 4, "Resilient Project Management Foundations," examines the fundamentals of project management in the context of the growth mindset framework. It begins with an overview of how standards are evolving to include resilience and flexibility as core competencies. In addition, projects must be aligned with company objectives and value delivery models. This chapter also explains fundamental project components, such as scope, resources, and timelines, as well as strategies for dealing with uncertainty and ambiguity. Finally, the chapter covers how using metrics to gauge performance provides a means to identify growth and improvement in response to shifting project needs.

Chapter 5, "Building a Dynamic Project Management Framework," explains how to create a dynamic project management framework based on predictive and adaptive project management strategies. The predictive (Waterfall) approach is explored in depth, presenting its phases and use cases for projects with specific, well-defined requirements. The Agile methodology—an adaptive approach—is analyzed from the growth mindset perspective, emphasizing flexibility and iterative progress. The dynamic Agile hybrid, which incorporates elements of both Waterfall and Agile, is also outlined as a useful method for larger projects.

Chapter 6, "The Growth Mindset in Stakeholder Management," outlines one of the most important parts of any project and explains how a growth mindset can boost engagement and cooperation with stakeholders. Building trust is at the forefront in this context, as this allows project managers to develop better relationships and promote collective ownership. The chapter also offers some tips for maneuvering into win-win situations with stakeholders to inspire their involvement in the project's growth and success.

Chapter 7 discusses how AI can help change the way projects are managed by improving workflow, decision-making, and a project manager's career development. The chapter embraces AI as a new global power in terms of efficiency, flexibility, and innovation that can also be mapped onto the growth mindset framework. The chapter considers AI-based predictive analytics and different tools for optimizing project management processes. It ends by sharing how AI enables project managers to shift their focus onto strategic work, challenge themselves, and grow professionally, unlocking new possibilities for collaboration.

Chapter 8, "Systems for Continuous Improvement in Project Management," examines the different processes and methodologies available to continuously improve projects. Building on the methodologies discussed in Chapter 5, it engages with systems such as Scrum, Lean, and Six Sigma, which support project managers in their growth through feedback loops and process optimization. Scrum's emphasis on short, focused Sprints, in combination with a growth mindset, enables project managers to be flexible and continuously learn. The Lean approach, including techniques such as Value Stream Mapping, emphasizes automation through the elimination of waste, while Six Sigma offers a data-driven path to excellence. The chapter also covers Kaizen, PDCA cycles, and the importance of instilling standardization to aid with scaling.

Chapter 9, "Product-Driven Strategies for Advancing Learning in Projects," concludes the book with a discussion on how to encourage innovation and learning within the context of projects by utilizing techniques originating in product management practices. It explores experiments and pilots for testing ideas and validating assumptions using data from beta testing and MVPs. By incorporating CX and UX research into a project's architecture, products can better meet real needs. The chapter also examines how demos can be effective for communication and validation, as well as how communities can be helpful for feedback and cooperation. With these product practices, project teams can continuously innovate and grow.

Chapter 1

Understanding the Growth Mindset

The concept of the growth mindset was first introduced in 2006 by Stanford professor Carol Dweck in her book *Mindset: The New Psychology of Success*.[6] Dweck's research demonstrates that, with a growth mindset, students are more likely to push themselves and believe they can do better, as well as be more capable, adaptive, and creative problem solvers. Dweck discovered that one of the most significant factors in student academic perseverance and progress is related to the way they think about intelligence. Students might view intelligence as a definite quantity that they either have or do not have (a fixed mindset) or, alternatively, as a variable quantity that can be further increased with practice and instruction (a growth mindset).

Fixed-mindset students assume they possess insignificant amounts of intellectual capacity and tend to worry about demonstrating rather than refining their intelligence. In the face of struggles and failures, they tend to experience negative thoughts, resulting in either feeling defeated or unworthy or simply giving up. Conversely, students with a growth mindset tend to accept failure as a learning opportunity. Consequently, they respond with positive ideas (e.g., "Perhaps I need to shift my strategy or try harder"), emotions (e.g., the excitement of the challenge), and actions (e.g., doggedness) in the face of the same adversity.

Fixed and Growth Mindset Characteristics

The concepts of a fixed mindset and a growth mindset were popularized as a result of years of research on thousands of children by Dr. Dweck. To have a growth mindset means understanding that our intelligence and capabilities are malleable with effort. This mindset empowers us to see problems as beneficial experiences and failures as stepping stones. Dweck did a study in which she recorded students' brain activity as they jotted down test errors. Interestingly, her notes reported no brain activity in students with a fixed mindset during this process, whereas those categorized as having a growth mindset showed active engagement.

Dweck found that while individuals with a fixed mindset consider their skills and abilities static and irrevocable, those with a growth mindset view intelligence and capabilities as attributes that can be learned, trained, practiced, and improved. This insight prompted a dichotomous distinction between the two ways of thinking. Fixed-minded individuals generally spurn difficult situations and are easily frustrated; they ignore or avoid feedback and feel threatened by what others have accomplished. In contrast, growth-oriented minds are open to challenges and bounce back from failure; they view effort as essential to mastery, see criticism as a learning opportunity, and emulate the success of others. The characteristics of the growth and fixed mindsets are presented in the table below, together with the triggers that may foster them.

Table 1. Fixed and Growth Mindset Characteristics

Triggers	Fixed Mindset	Growth Mindset
Challenges	Avoid	Embrace
Obstacles	Give up easily	Persist in the face of setbacks
Effort	See it as fruitless	See it as the path to mastery
Criticism	Ignore or avoid	Learn from it
Success of others	Feel threatened	Find lessons and inspirations

The results of having a fixed mindset can be detrimental by reducing opportunities for learning and development, as well as causing underachievement, broken relationships, limited career opportunities, and a low sense of self-fulfillment. By contrast, a growth mindset has a number of demonstrated benefits, such as improved resilience, tolerance, learning ability, problem-solving capacity, perseverance, willingness to receive feedback, self-image, flexibility, relationships, career progression, and general outlook on life.

The table below shows examples of the effects of having either a growth or a fixed mindset.

Table 2. The Effects of a Fixed and Growth Mindset

Examples of the Effects of Fixed and Growth Mindset	
Fixed Mindset	**Growth Mindset**
Limiting learning and growth	Increased resilience
Underachievement	Embracing challenges
Strained relationships	Enhanced learning
Limited career advancement	Improved problem solving

The Evolution of the Growth Mindset Theory

Over years of studying this subject, Dweck found it important to clarify that a growth mindset is often mistaken for merely making an effort. It is designed to bridge achievement gaps, she explained, not obscure them, and to acknowledge a student's current successes while working with them to develop their potential. Mary C. Murphy, a professor of Psychological and Brain Sciences at Indiana University who collaborated with Carol Dweck, later extended this theory, demonstrating the potential irony inherent in the attempt to label mindsets in a binary fashion.[7] This way of thinking can also lead to moralizing individuals and equating a growth-oriented mindset with being a better person. Such terms overlook the context and culture in which mindsets are developed and shaped, rendering their use in recruitment strategies and the assessment of organizations more complicated.

Both Murphy's and Dweck's studies led to the conclusion that no person is unique in having a fixed or growth mindset. Individuals may lean toward one or the other, but everyone is capable of thinking one way and subsequently changing to the other. It is more accurate to consider mindsets in terms of a continuum, and thus an individual's place on this continuum at any given time is governed by the context of the situation or the institutions involved. Our own status in this context is worth monitoring in order to recognize when it is that we become stuck in the fixed mode. If the context is the growth mindset in terms of project management, as this book aims to elucidate in detail, it is essential for project managers to become self-aware in this sense, understanding their own mindset habits, as well as those of their team members and the company culture. Armed with this knowledge, project managers can respond to obstacles and threats, as well as prepare themselves for all sorts of scenarios.

Adoption of the Growth Mindset in the Corporate World

In 2010, Dweck and three colleagues—Mary C. Murphy, Jennifer Chatman, and Laura Kray—began working with the consulting firm Senn Delaney to explore the role of mindset at a corporate level. They surveyed a diverse set of employees across seven Fortune 1000 companies and asked them how much they agreed with various statements, such as "Successful people in this company seem to believe talent is a fixed thing that can't be changed much." High agreement with such questions meant the organization's mindset was most likely fixed, while low agreement scores represented a growth organizational mindset. The team then assessed organizational mindset in terms of the growth/fixed dichotomy, focusing especially on the impact that its manifestation in one way or another had on employees' satisfaction, perceptions of organizational culture, collaboration norms, innovative behavior, and ethical conduct, as well as supervisors' perceptions of their employees through a survey.[8]

According to Dweck, companies often exhibit a strong consensus on mindset at an organizational level. However, this overall alignment is naturally complemented by individual variations, as employees bring their own unique mindset traits to the table. For example, companies with a fixed mindset only value "star" employees, so only a few are considered important, leading to lower company loyalty. They are terrified of failing; as a result, they resist innovative projects and sometimes resort to secretive or deceptive practices to succeed. In growth-mindset companies, in contrast, supervisors see employees primarily as innovative and eager to learn each day. They contribute to a greater recognition of leadership capabilities within their teams as well.

Dweck's work has been widely adopted by the corporate world, which understands that a growth mindset is a catalyst for good company culture and performance.[9] It took companies little time to appreciate the value of a growth mindset in business. Corporations realized that continuous learning is crucial because it is through their willingness to acquire and adapt to new information that they can navigate a dynamic marketplace. A growth mindset implies being open to new possibilities, unlearning old patterns, and experimenting with new approaches, all of which are essential for personal and professional growth (and, therefore, for organizational development).

The growth mindset philosophy aligns seamlessly with the competitive nature of the business world, in which the desire to continuously strive to be better is just one of the many aspects that company leaders increasingly emphasize. Corporate leaders push employees to find new opportunities and venture outside of their comfort zones, even when it seems risky. Business requires quick-thinking employees who are willing to make big decisions without the worry of making mistakes. These employees treat mistakes not as failures but as learning experiences that enable them to bounce back quickly, identify what went wrong, and rectify the error.

Research shows that growth-minded individuals can be better leaders because they see the potential in others rather than dwelling on their own flaws.[10] Such leaders believe that humans can change and develop, resulting in more positive thinking and making better choices. Furthermore, the cultivation of a growth mindset is increasingly perceived as an essential aspect of employee retention and success. While most organizations view it as an issue of personal motivation and growth, the data suggests that a growth mindset also encourages an organizational culture of learning. This type of culture emphasizes re-skilling and upskilling.

When the 2021 "Great Resignation" took place, displacing more than 30 million Americans, organizational behaviorists noted that people not only left their positions, they also left dysfunctional work cultures. Growth-minded individuals see their work as a way to make a difference, not just as a tool to obtain money. This perspective can be motivating, especially for those who feel a lack of control in their lives. A major area where companies can demonstrate a fixed or growth mindset is in their people management practices.

According to Murphy's research, Patagonia, an American outdoor apparel and equipment company, is an example of an organization that eliminated individualized performance ratings altogether. Dean Carter, Patagonia's Human Resources Director, understood that the brand was so interested in what it could get out of individuals that it never cared about what it gave back. He noticed that employees were exhausted by the review process, and the company offered raises and bonuses to boost morale. Instead of yearly reviews, their HR department now provides employees with a resource that they choose to use to help them improve their performance—depending on their situation and their manager—rather than such a resource being forced upon them by HR. This new practice has increased employee engagement and provided more insight and data to help with further growth. In addition to providing HR with new performance analysis techniques, HR is now also provided with much more useful information. Carter further noted that the system allows both employees and managers to concentrate on their tasks more effectively.

Murphy also offers the example of the Talent Assessment Program at GitLab, a software company that encourages managers to examine not only performance but also ways in which their team can develop. According to the company, growth potential is defined as a person's willingness and capacity to perform more complex or wider tasks and learn new skills compared to their peers and roles; this includes the possibility of progressing to other levels. The more growth potential an employee has, the more managers can help them identify the best route to fulfilling that potential, recognizing always that this potential can change over time as their skill set, talents, and interests evolve as well.

Another example Murphy provides is a description of her experience with the executive vice president of HR at Shell, Jorrit van der Togt. In 2007, Goal Zero was Shell's vision for ultimate safety: zero injury and zero leaks across the entire organization, including personnel, process, and transport. Shell failed to achieve Goal Zero despite tireless efforts. Although their safety numbers had climbed steadily, they wondered if a growth mindset culture could help close the gap. This brought up the question of whether employees could break away from fixed thinking and habits and commit to learning—even when mistakes have been made.

While Shell was developing new commercial methods to meet an uncertain future, van der Togt was revising the company's talent management program. He sought inspiration at Stanford, where

Murphy demonstrated her paradigm shift in terms of mindset culture. After her talk, he was encouraged to use as much creativity as possible to approach the problem. Shell's changing of its business model and reorientation of its mindset came at a timely moment, as the need for innovation in its sector increased even more than in others at that time.

Chapter 2

Golden Rules of the Growth Mindset in Project Management

The golden rules of the growth mindset in the context of project management represent a set of characteristics that define an ideal growth mindset approach. Like any model, this is a generalization used in this book to highlight overarching trends, illustrate phenomena, and provide a framework for selecting specific examples. As mentioned earlier, every individual—including project managers—embodies a combination of fixed and growth mindsets, responding to various conditions shaped by their external environments, such as team dynamics and company culture.

Figure 1. Ten Golden Rules of the Growth Mindset in Project Management

TEN GOLDEN RULES OF THE GROWTH MINDSET IN PROJECT MANAGEMENT

1. Continuous Improvement
2. Learning from Feedback
3. Creating a Collaborative Culture
4. Fostering Innovation
5. Taking Risks Without Fearing Failure
6. Thriving in Ambiguity
7. Embracing Challenges as Opportunities
8. Bouncing Back from Setbacks With Resilience
9. Facilitating Knowledge Transfer
10. Leading by Example

Continuous Improvement

Continuous improvement refers to the ongoing endeavor to improve every aspect of an organization. It is based on the idea that a relentless flow of modifications will, when carefully applied, deliver game-changing results. Continuous improvement isn't just a way of doing things—it's a way of thinking that helps an organization to continually evolve. It has its foundations in the belief that small, incremental improvements, if made with purpose, will make a difference. It is about seeing obstacles and failures as a path to learning, not as limitations.

Encouraging continuous improvement through the adoption of a growth mindset starts with behavior—fostering long-term changes in which employees embrace new ways of working and learn to act differently. This shift demands that we perceive obstacles and changes as opportunities, not threats. Rather than aiming for large, disruptive changes, minor incremental improvements often yield the most sustainable results. This fits well with the growth mindset theory, which asserts that abilities and results are improved through practice and effort. Over time, small changes and tweaks lead to meaningful progress. It is also much easier to spread these little successes across the organization and reinforce the notion that there is always more growth to come for all.

While the concept of continuous improvement is definitely a component of the growth mindset, the flip side of this emphasis on constantly maximizing performance is that it also promotes perfectionism, which is a fixed mindset characteristic. In today's world of fast-paced work, it is not always a question of perfection but of growth. As a corporate saying goes, "perfectionism is the worst quality of work," since the pursuit of perfection can limit risk-taking and experimentation. Experimentation is at the heart of continuous improvement, where employees are invited to test out new ways of working. The growth mindset thrives in environments where failure is experienced as a teaching moment rather than as a failure.

Ultimately, continuous improvement for a project manager entails the belief that each project provides something new to learn and improve on. This implies a willingness to obtain new skills, try new project management tools, or incorporate diverse opinions into the decision-making process. Project management evaluation tools, such as performing post-mortems or documenting lessons learned and best practices, greatly serve this purpose. They allow for the monitoring of progress in terms of continuous improvement and the calculation of the learning value of the project, as well as providing insight into patterns and trends. Measured improvement ensures that progress is visible and tracked. Regular reporting keeps the process on track and organizations moving forward; however, in the context of a growth mindset, the act of measuring is about understanding how learning and effort lead to growth. Regularly evaluating progress reinforces the belief that improvement is possible for everyone, and sustaining this process over time ensures that growth continues.

The paradigm of the learning organization takes this a step further by highlighting how organizations need to be competent at adapting their actions in response to new information and insights. This creates a mindset in which people are constantly learning, developing, and disseminating what they are learning, which is a cornerstone of innovation. Importantly, this includes learning from the successes of others. The saying "success leaves clues" is certainly true when it comes to project management, for it suggests that success is not luck—it is a product of specific actions and behaviors, which are observable and can be reproduced. However, this does not mean it is an easy task. The success of others can become a source of anxiety, especially in the context of a competitive company culture. Even in helpful, less hostile environments, learning about the accomplishments of other teams takes effort. Gaining insights from winners can even be transformed into a team ritual if it happens through such methods as workshops, collective case studies, and having winning teams share their stories.

Finally, industry examples and evidence may enable teams to develop both collaboratively and creatively and to learn from other companies' successes.

Learning from Feedback

A crucial learning tool that is easily accessible to a project manager is feedback. Feedback tells a project manager what is working well and what needs to be changed, as well as how to revise it. A good growth mindset practice is to proactively ask for feedback even before the project's completion, as this may help to make necessary adjustments and mitigate risks before it is too late.

Despite understanding that feedback facilitates our growth, it is nevertheless natural to feel nervous about receiving it. The growth mindset way of managing this situation would be to control one's emotions and do something productive instead. It may also be helpful to take a moment after receiving feedback to consider it—especially if the messages are constructive or challenging—in that this pause might prevent an impulsive reaction. Stepping back to take a broader view can be useful as well; it is important to bear in mind that feedback should always be related to the performance of tasks and not to a person and that everyone in the project or organization should give and receive feedback.

Responding well to feedback is certainly an aspect of a growth mindset and is related to the faith that we can increase our understanding (and outcomes) over time with effort and determination. Cultivating a growth mindset involves mindfulness, self-observation, and, ultimately, greater awareness. The best way to learn this skill is by incorporating feedback and working towards an understanding of how things could have been done differently. One issue that can arise is that the received feedback is too ambiguous. If faced with this situation, asking questions or using examples can often help to clarify misunderstandings and thereby reinforce a growth mindset. In addition, the steps taken after receiving feedback are just as important. Taking a proactive strategy, openly sharing feedback with the team, and coming up with ideas for improvement demonstrate a growth mindset approach, which can be both helpful for a project and inspiring for others.

Customer feedback is its own distinct and crucial topic, as it is hard to imagine driving innovation without it. Customer feedback can be extremely valuable for companies that want to improve their product development cycle. Furthermore, incorporating customer feedback into the product innovation lifecycle can create a feedback loop that builds strong consumer relationships. When customers realize their voices are heard and acted upon, they will be loyal and trust the brand to cultivate long-term relationships. Businesses adopting this co-creative spirit can better respond to shifting market dynamics and consumer demands. By engaging with customers through surveys, social media, and direct conversations, companies can collect valuable data that reveals the needs and preferences that can lead to disruptive innovations. This is exactly what Coca-Cola did to leverage customer feedback in its product development. Since launching its touchscreen soda fountain, the Freestyle Coca-Cola has been collecting massive amounts of customer insights.[11]

Leveraging Customer Feedback for Product Development: The Coca-Cola Freestyle Experience Case Study

Coca-Cola Freestyle is an AI-enabled, cloud-connected touchscreen soda fountain offering a choice of over 165 flavors, including the option to mix different products. In 2022, more than 50,000 Coca-Cola Freestyle machines were in use globally, dispensing over 11 million drinks daily.[12] Customers can 'pre-order' their drinks, pay through the app, and pick up their beverages at the nearest location. This capability provides Coca-Cola with valuable insights into its customers' preferences.

Users must sign up with their social media profiles to access the mobile app. Coca-Cola leverages AI to analyze data from users' social media feeds, gaining insights into where, when, and how customers consume its products. Based on consumer behavior and demographic analysis, Coca-Cola identifies the most popular products in specific regions. This data is then used to create targeted marketing messages. Because the machines are cloud-connected, beverage prices can be remotely adjusted based on promotions available at specific outlets.

With the ability to track performance in real-time and display customizable content, Freestyle screens can be localized, personalized, and synchronized with larger campaigns. Coca-Cola's system integrates with Adobe Experience Manager, enabling the creation of personalized shopping experiences at scale and driving significant increases in customer engagement and revenue. Marketers at Coca-Cola can upload campaign assets and soda brand collateral into the system, setting rules to determine which machines display specific content tailored by campaigns, regions, or customer profile.

AI-equipped vending machines represent a significant breakthrough for Coca-Cola in market research. Through continuous product performance monitoring, the company has developed robust market intelligence to stay ahead of industry trends. First introduced in 2009, the Freestyle project remains ongoing, with new technical features regularly added.

Felicia Hale, Vice President of Equipment Strategy for Coca-Cola Freestyle, stated in a press interview: "We're finding that when a consumer goes up to our Freestyle, they're almost looking for what's new and different, and that newness and freshness really helps to inform our pipeline and how we think about innovation."[13]

Since feedback is a broad category, this topic is explored further in Chapter 6, "The Growth Mindset in Stakeholder Management," in the *Continuous Feedback Loops* and *Types of Feedback* sections.

Creating a Collaborative Culture

A growth-minded project manager contributes to developing a growth-oriented culture of collaboration at both the team and organizational levels. Specifically, they establish a culture that encourages building connections, partnering with others to complete tasks, and promoting the values of effective project knowledge sharing. This often comes with a positive and encouraging attitude, enabling the project team and key stakeholders to 'buy in' and actively engage in project management efforts.

The Project Management Institute highlights the act of creating a collaborative project team environment as one of the project management principles, defining it as a cooperative project team culture that enables alignment with other organizational cultures and standards, facilitates learning and development for both the individual and the group, and promotes the best effort to achieve the desired result.[14] Many different aspects contribute to the creation of a collaborative project environment, including agreements among teams, frameworks, and workflows. These elements together create a culture that allows people to cooperate and generate mutually beneficial effects from exchange.

The structures used, designed, and implemented by project teams facilitate the individual work involved in project activities. Organizational architectures refer to any configuration or linkage between aspects of project work and organizational development. They are defined and tailored to the context of the project or designed entirely for a specific project use case. Team agreements are a set of behavioral rules and working practices developed by the project team and maintained through commitment at both the personal and project team levels. The team agreement should be established at the outset of a project and will evolve as the project team works together and learns about what norms and behaviors are needed for the team to continue to work in a collaborative and productive fashion.

Team dynamics on projects are driven by the culture of the project organizations, the project context, and the space in which teams operate. Salesforce serves as a prime example of an effective collaborative culture. Salesforce is a cloud-based software company offering CRM solutions and enterprise applications to help businesses manage customer relationships. It was founded by entrepreneur Marc Benioff in 1999. Since Marc Benioff's time, Salesforce has emphasized collaboration as its key to success. Benioff's "Ohana," Hawaiian for family, is a concept that promotes openness, honesty, and collaboration where employees, customers, and partners innovate together. One of the company's most successful projects is its focus on cross-functional collaboration for generating customer-specific solutions. Sales, product, customer service, and marketing teams collaborate frequently to ensure that the company's products are compliant with the market and provide seamless user experiences. This sharing strategy is responsible for Salesforce's success with its CRM platform, which has grown by integrating different services. By creating a culture of shared responsibility and transparency, Salesforce not only revolutionized CRM but is now one of the most prominent cloud computing and enterprise platforms.

Fostering Innovation

Innovations are crucial to project management as they help teams overcome challenges, evolve, and achieve outcomes beyond expectations. Innovation keeps projects competitive in a dynamic business world where everything moves quickly. It elevates the everyday execution of projects to a strategic activity with distinctive added value for the stakeholders and the organization itself.

A growth mindset is the basis for implementing innovations. It enables the project manager and the team to see obstacles as an opportunity to improve, not as a threat to performance. This mentality evokes curiosity, resilience, and an openness to failure, all of which are vital for creativity. Growth mindset project managers will find new ways to see, re-think, and try things out—and their teams will follow suit. This process implies a move from simply getting things done to constantly striving for better and more efficient ways to achieve goals.

In terms of innovation, project managers must create a space in which creativity can thrive. In such a space, teams can innovate and move outside the box. The critical factor in this case is psychological safety, which implies that team members are comfortable voicing unconventional thoughts without fearing being dismissed or disappointed. By granting them the resources (time, tools, knowledge) that they need, the team can experiment and make mistakes. Additionally, stretch goals can inspire the team to think outside the comfort zone of their daily work and pursue original ideas.

Lastly, the project manager can be an innovator as well. They experiment with ideas, implement new methodologies, and optimize processes by applying ongoing improvements and flexibility. Whether this means using the most innovative project management software, trying new workflows, or taking inspiration from other projects, their pursuit of innovation should become an inspiration to the team. This provides a source of confidence and also encourages the whole team to embrace experimentation as a route to growth. By doing this, the project manager will not only be a source of innovation but also a creator of it because their projects deliver not only results but also valuable innovation.

Throughout history, many innovators have demonstrated remarkable perseverance in the face of failure. Thomas Edison's success with the light bulb did not happen overnight; it was preceded by a series of failures. He overcame approximately 2000 failed experiments to create what we now recognize as the modern light bulb. Moreover, we know that he did not invent the electric light—it was Davey in 1802, and there were actually more inventors involved in this process.[15] In 1840, De la Rue introduced platinum coils in a vacuum tube. Swan took it from there and placed the carbonized paper into a glass bulb. Following that, Woodward and Evans secured the first patent for an electric lamp, which was later acquired by Edison. Edison's dedication to fostering innovation came together with his eagerness for continuous improvement, his risk-taking approach, and his ability to bounce back from setbacks. At the conclusion of his journey, he famously stated: "I have not failed. I've just found 10,000 ways that won't work." This is a perfect exemplification of a growth mindset.[16]

Edison's words are great ones to revisit at this moment, when innovation has become the highest priority for most organizations. Looking at the evolution of technology companies, we see that technological progress and many other fields are only possible when we seek out challenges and fail (many times), reflecting on these failures to build upon them later. One of the innovators who bounced back from setbacks with resilience is Chester Carlson, the founder of the Xerox Corporation. Chester Carlson spent years perfecting his creation while working a day job and attending law school during the evenings. In addition to all that, he was conducting chemical experiments in a kitchen, which drove his wife crazy. She eventually forced him to move his office to a second-floor room in a house owned by his mother-in-law. Eventually, though, after more than 10 years of struggle, he teamed up with the Haloid Company, whose product, though superior, cost nearly 10 times as much as other machines.

Despite appeals to major players such as Kodak, IBM, and GE, none responded with interest, reasoning that they could not see the value enough to want to pay for it, at least not until Joe Wilson, Haloid's president, came up with a billion-dollar idea: instead of selling the machines themselves, why not lease them? This approach worked, transforming into the company we know today as the Xerox Corporation.[17] Chester Carlson wasn't a project manager per se, but he certainly managed many projects and successfully bounced back from setbacks many times. The ability of the project manager to pivot and change approach will affect where that project will end up. This is why being able to bring a project back in line when it gets out of control should be considered a part of the project manager's responsibility. Similarly, collecting tools, best practices, and strategies for recovering from setbacks should be part of the project manager's continuous learning practice.

Taking Risks Without Fearing Failure

A good starting point in terms of risk assessment is to determine what risk-taking means to the project team, the company, and with respect to its mission. Risk is not synonymous with impropriety, irresponsibility, or carelessness. Instead, it is about new perspectives, reversing presumptions, and experimenting with hypotheses to improve things. Good risk-taking aligns with the team's vision, values, and mission and is, moreover, driven by data, evidence, and customer feedback. A growth mindset project manager can help the team understand the benefits and downsides of taking risks by setting realistic expectations, giving examples, and explaining how and why.

To be able to take risks in experimentation without the fear of failing, it is also important to define success and failure. Apart from the company's project success metrics, it is equally important to understand in depth what success means for the project manager's own process of self-discovery and for the team's growth. This not only makes the evaluation processes rich in terms of attaining more data points, but it also triggers a positive impulse to keep people engaged. It promotes an "I care" attitude, which, after all, is something the whole project will benefit from.

SpaceX suffered three unsuccessful launches between 2006 and 2008, nearly going bankrupt at each attempt. Rather than give up, SpaceX accepted its own mistakes and launched its fourth Falcon 1 in 2008, making it the first private-money rocket to attain orbit. This success earned SpaceX a $1.6 billion contract for supply flights to the International Space Station from NASA. This success stemmed from promoting an environment of iterative development, viewing failure as the pathway to creativity. SpaceX is now at the forefront of reusable rocketry and interplanetary explorers.

Curiosity is key because it is the fuel for risk. It is curiosity that gives humans the freedom to continuously seek out new knowledge, insights, and solutions. Curiosity management strategies include the following:

asking questions that are open to all, allowing for diverse feedback and input, and recognizing and rewarding curiosity. In addition, providing the opportunity to learn from various experts, clients, or peers and communicating it to the team can help build a culture of exploration. As a project progresses, taking risks and showing the team a risk-taking approach, sharing a project manager's thoughts, experiments, and failures freely and openly, as well as showing vulnerability and humility, may inspire everyone to leave their comfort zone, take on new activities, and learn new techniques.

As a contrary example, in the case of BlackBerry presented below, we can see just how costly it can be to avoid taking risks, thereby missing the opportunity to introduce innovation early.

How the Fixed Mindset Limited Growth at BlackBerry: A Case Study

BlackBerry dominated the smartphone market from the early 2000s until about 2008. It ruled the business and enterprise worlds during this period, mostly thanks to its secure email service and legendary physical keyboard, which professionals used for fast typing. BlackBerry handsets like the BlackBerry Bold and Curve phones were considered a communication must-have device in corporate contexts. The company was also a perfect example of how a fixed mindset would crush innovation and make a company go bankrupt.

BlackBerry first built a fortune by providing safe, business-grade communication products and became the smartphone of choice for business users thanks to its traditional keyboard and email offerings. However, as customers turned to multipurpose touchscreen smartphones, BlackBerry's top management remained firmly entrenched in the virtues of its physical keyboard and closed platform. They ignored the burgeoning desire for touchscreens and mobile apps as pointless advancements, maintaining their focus on the enterprise sector while underestimating the needs of the consumer market.

That unwillingness to innovate and evolve resulted in several critical errors. When Apple first launched the iPhone in 2007, for example, BlackBerry executives branded it as a "niche" product that couldn't compete with the business-oriented BlackBerry phones. The firm also lost out on the chance to create its own app store, which has become an integral aspect of the iPhone and Android ecosystems. When BlackBerry finally made the decision to make the switch and launch touchscreen devices such as the BlackBerry Storm, the market was already overtaken by iPhones and Android phones, which were far more flexible.

BlackBerry's inability to take risks and shift from a status quo mentality to a growth mindset that would allow it to incorporate new technologies and customer demands was a major factor behind its decline. As much as the brand had been reenergized with new software and partnerships, its refusal to make radical shifts, particularly on its hardware and interface, inevitably meant it was unable to keep up in a more innovation-centric marketplace.

Thriving in Ambiguity

While often uncomfortable and difficult to handle, ambiguity is essential to pursuing new avenues of possibility while figuring out the path forward. The late Bob Galvin, longtime CEO of Motorola, often spoke of how to harness the power of creative thinking to overcome difficult challenges. He believed that uncertainty and relentless inventiveness were the keys to discovering remarkable solutions, rendering decision-making easier and more efficient.[18]

The COVID-19 pandemic has been a challenging time in multiple dimensions—both personal and professional—for companies and employees on programs, projects, and processes. Project managers, irrespective of the type of organization they work for—from small businesses to large corporations—and the geographical locations they are based in, face multiple roadblocks.

In large organizations, the sudden shift to remote working disrupted well-established workflows and required new tools for communication and collaboration, leading to delays in project timelines, increased budgets, and problems with resource allocation. The disruption to supply chains led to delays and increased costs of certain project inputs. Many had to redefine the scope of projects due to budget constraints. Stakeholders had to re-align on priorities more frequently than before, and project managers had to frequently communicate with them to understand and address their changing expectations. The pandemic created a highly stressful environment, and project managers had to be attentive to the mental health of employees and support their wellbeing. For small businesses, the sudden shift to remote working has been a challenge due to limited resources to absorb these disruptions. Cash flow issues have created problems with funding projects and retaining employees.

This has also made maintaining stakeholder and client relationships more difficult. Business pivots rendered some needs for the end project redundant and revealed new skills that teams needed in terms of digital tools. In-person work was complicated by health and safety considerations. In both contexts, project managers had to grapple with similar problems: how to manage uncertainty and risk, how to overcome communication issues in remote work settings, and how to quickly learn and adopt technology for these new project-management environments.

Slack Technologies showed how project management with a growth mindset can enable a business to thrive during times of uncertainty like the COVID-19 pandemic. With the spread of remote working, Slack jumped on the bandwagon and used Agile to change the direction of its product development. Project managers switched gears and released enhancements such as integrated video conferencing and virtual workspace tools. This process was based on iterative development and a quick response to feedback. Slack's constant evolution based on real-time feedback produced a spike in active users—by the close of 2020, there were 12 million daily active users versus 10 million at the start of 2020.[19]

This company's success shows how project management requires transparency, collaboration, and flexibility. Slack's teams spent the entire time connecting with their users, gathering feedback to help inform product development and ensure that new features were developed in response to real-world issues. This approach resulted in higher customer satisfaction and drove explosive growth—in FY2021, Slack's revenue increased by 43% to $902 million. Through customer engagement and relentless iteration, Slack tapped into the growth mindset and made uncertainty a catalyst for change and continued success.[20]

Embracing Challenges as Opportunities

Reframing challenges as positive opportunities enables project managers to adapt and push beyond their comfort zones to eventually overcome any hurdle blocking the path to the fulfillment of their role. The history of Netflix illustrates how embracing challenges as opportunities at a company level can drive success.[21] In 1997, Reed Hastings and Marc Randolph came up with the idea of renting out DVDs by mail. They started really small, doing everything by themselves. A year later, they launched Netflix.com and started offering consumers a collection of DVDs available for rental. Being eager to grow, the following year, in 1999, the company added a subscription service—highly innovative at the time—that allowed for unlimited DVD rentals.

Despite observable growth, the challenge at the time was due to the fact that they remained in the shadow of the dominant Blockbuster. Moreover, this was all happening during the advent of digital streaming. Rather than slowly pushing forward with the trends, Netflix found a growth mindset solution. They saw this as an opportunity to reinvent itself, which they did by investing a large amount of money in streaming technology and programming. This shift was a huge risk given the cost of investment, the potential loss of loyal customers, and direct competition with established networks and studios. However, today, we know that this tale has a happy ending. Netflix became a global streaming leader, changing the way people consume media. Through its risky move, Netflix was able to break into new markets, attracting a much wider global audience and setting a new standard in entertainment.

Project managers can learn the lesson from Netflix that in challenging situations, instead of viewing them as barriers, we should find ways to turn them into opportunities for innovation and growth. A growth mindset enables project managers to take calculated risks and lead their teams through disruption. In this way, they can motivate sustained achievement and transform setbacks into game-changing successes.

Bouncing Back from Setbacks with Resilience

Project resilience means a company or project team can change, recover, and remain stable in the face of unanticipated risks and disturbances. Project resilience is the capacity to prepare for and react quickly to obstacles and delays that could threaten project execution. It requires planning, adaptability, collaboration, and continual improvement.

Bouncing back from setbacks is actually a part of the company, product, and project cycle. In 2011, LEGO faced a challenging decline in sales. To address this issue in a transparent way, the company launched the LEGO Ideas website, where users could design their own sets and vote on which ones they liked best. Thanks to that innovation, LEGO could access its communities' creativity directly and transform potential challenges into a force for innovation. The well-known triumph of this project was the Women of NASA set, which was developed based on fan feedback, as well as others, such as the NASA Apollo Saturn V set. Today, LEGO is the market leader in toy production, beating out its competitors and securing a significant market share. Despite an inflation-driven sales decline for the toy market in the first half of 2024, LEGO's revenues jumped 13% to 31 billion Danish kroner (about $4.65 billion). They bounced back from a crisis even stronger than before, ready to drive further growth.[22]

A project manager may not have a direct influence on company policies, but they can play a key role in fostering a culture of resilience. The Project Management Institute (PMI) identifies embracing acceptability and resiliency as one of the project management principles, defining it as the capability to "build adaptability and resiliency into the organization's and project team's approaches to help the project accommodate change, recover from setbacks, and advance the work of the project."[23]

A project rarely succeeds as initially intended. Projects are influenced by internal and external factors, new needs, problems, stakeholder engagement, etc., which exist in a dynamic exchange framework. Adaptation must be holistic, including an appropriate change control process, in order to minimize scope creep. According to The Project Management Institute, adaptability and resilience in a project include the following:

- Quick feedback loops
- Continuous learning and improvement
- Well-rounded project teams with a large skill set
- Regular inspection and adaptation of project work
- Multiple project teams to capture a broad range of experiences
- Transparent and open planning across both internal and external partners
- Prototypes and experiments on a smaller scale to test ideas and experiments
- The ability to apply new ways of thinking and working
- Process planning for both work speed and requirements stability

- Open organizational conversations
- Multidisciplinary project teams of diverse skill sets, cultures, and experience
- Learning from earlier studies or the same activities
- Flexibility and ability to anticipate different possibilities and plan for different events
- Deferring decision-making to the last responsible moment
- Management support
- Open-ended design that balances speed and stability

Following PMI's principles—concentrating more on envisioning outcomes rather than on deliverables only—the project manager can benefit the project and stakeholders, and the result may be better than expected. For example, a project team might discover better ways to deliver higher-quality results than the initially agreed-upon deliverable. Unforeseen changes and circumstances during a project may uncover new opportunities.

Facilitating Knowledge Transfer

Both knowledge management and knowledge transfer play critical roles in empowering the effective use of the growth mindset in project management by enabling learning from projects and avoiding the loss or waste of insights gained, lessons learned, and best practices achieved throughout a project's execution. By systematically capturing, storing, and sharing project knowledge, organizations can improve the performance, quality, and efficiency of their current projects, as well as facilitate the learning and development of project teams and stakeholders and enhance future projects' performance, quality, and efficiency.

General Electric (GE) is a Fortune 500 company with a long history of leadership development and knowledge transfer among its employees around the world. One of the most illustrative examples of GE's initiative in terms of knowledge transfer is the GE Management Development Institute at Crotonville, New York. The institute was set up in 1956 and has been at the heart of the company's plan to pass on valuable knowledge to its leaders. At Crotonville, GE gathers employees from across different regions and business areas to learn from one another, including best practices, and build leadership skills. By providing Leadership Training, Executive Development Programs, and Cross-Functional Training, GE promotes a culture of learning and sharing across the organization. For instance, as GE expanded into more technological sectors, it used knowledge transfer to successfully transition its leadership from traditional manufacturing to digital and industrial IoT (Internet of Things). GE leaders from all fields and geographies taught and learned from one another via formal training, mentorship, and joint projects.

Mentorship also played a significant role in GE's knowledge transfer strategy. The senior executives led new managers by advising them on their experiences, ideas, and knowledge related to key functions, including operations, market growth, and risk. This was especially important in times when leadership was changing, as new people could bring a culture of innovation and development to GE. The outputs of such knowledge exchange efforts were evident. GE navigated multiple transitions, including from manufacturing to technology and from local to global markets, without sacrificing its leadership pool. The company's flexibility across multiple markets, such as healthcare, energy, and aerospace, can be attributed to its standards for sharing key expertise and resourcing a strong and diverse workforce across its many divisions.

A core characteristic of the growth mindset project manager is self-awareness, which applies to all activities, including knowledge transfer. An essential mindful practice is to specify the type of knowledge to be captured before beginning the collection and dissemination process. Why is this knowledge needed, and who will benefit from it? A knowledge management strategy or plan can articulate objectives, scope, sources, methods, tools, roles, and responsibilities. A project manager's objectives should align with project goals and outcomes, as well as the needs and expectations of the project team and stakeholders. What is the type of knowledge (explicit versus implicit), the level of expertise (organizational, sector, project), and the format of knowledge (formal documents versus unstructured or narrative data)? This will help render the knowledge transfer process more structured and provide incentives for a growth mindset orientation.

Project management practices—in the form of a knowledge bank to hold the knowledge, a structured process to transfer it, and a knowledge network to drive it—can be used to support this objective. Project managers can demonstrate and educate co-workers about the advantages of a growth mindset approach and the aims of knowledge management activity. In this way, a project manager can invite suggestions from the project team and stakeholders regarding improving knowledge management and transfer processes.

Leading by Example

When millions of cars from General Motors were recalled in 2014 for having a faulty ignition switch that was the cause of dozens of accidents and deaths, the issue became quite dramatic. This episode represented a real threat to the leadership at GM, one that required immediate steps to restore trust and accountability within the corporation.

Mary Barra, the CEO, responded by acting in an open and personal manner. She took full responsibility for the crisis, rather than shifting blame, and turned herself into the catalyst of the company's transformation. Barra publicly admitted that GM made mistakes and would do what was suitable for the company and its stakeholders. In addition to that, she implemented the "Speak Up for Safety" program, where employees could voice safety concerns without fear of reprisal. This move demonstrated her insistence on putting safety and transparency at the center of GM's operations. Furthermore, Barra engaged customers, employees, regulators, and the public in frank conversations that reinforced a sense of trust and integrity in the company. GM emerged from the crisis as a victor, creating a culture of integrity and teamwork. Barra led by example, encouraging workers to emulate these principles and helping GM achieve a dramatic uptick in customer satisfaction and the restoration of the company's reputation.

Through her own examples, Barra showed how being a role model through transparency and responsibility can provide both immediate crisis management and long-term culture shifts. Similarly, one of the growth mindset project manager's characteristics is the ability to inspire co-workers to go above and beyond. The culture of model leadership builds a sense of shared ownership and drives others to deliver. The experience of seeing their leader approaching a project with a growth mindset creates a sense of meaning and urges their team members to contribute as well.

Growth mindset project management leadership does not imply going it alone. True leadership comes down to the behaviors and attitudes that project managers hope to inculcate in their workforce. When project managers adopt the growth mindset, they set an example for their teams that will help them be successful. By leading projects with a growth mindset and a clear dedication and willingness to take on challenges, project managers foster a culture of dedication within their team.

When team members are aware of their project manager's participation in the work, their incentive to deliver goes up. To lead by example creates a culture of collective responsibility and requires everyone to try their best. The real passion their leader puts into the project creates a sense of purpose for the team, and everyone has an urge to help. This collective drive generates a focused, agile team that can overcome challenges and meet project goals efficiently.

A project manager who leads by example opens the doors to creativity and expansion. They motivate team members to think outside the box and look for ways to do things better by getting involved in project activities, embracing fresh perspectives, and fostering a culture of continuous learning. This approach enables creativity in finding solutions, efficiencies, and cutting-edge concepts that move the project forward, as well as enables the team to overcome obstacles and react to shifting project requirements.

Chapter 3

The Growth Mindset Ecosystem

Building a growth mindset ecosystem involves multiple layers, including organizational and leadership-related aspects, the project team's collective approach, and finally, addressing the personal perspective of each team member. All of these tiers interact with each other dynamically in ways that can support either a growth or fixed mindset culture.

Figure 2. Growth Mindset Ecosystem

There are many ways the dynamics among the different elements of the ecosystem can interact. The fixed mindset culture can function at both a group and an organizational level, and its influence can be strong enough to obstruct an individual's ability to adopt a growth mindset. However, it can also work the other way around: a leader with a strong growth mindset can lead by example, inspiring more and more colleagues to adopt a growth mindset approach, ultimately promoting a growth mindset culture within the organization. There are many lines and directions through which different dynamics unfold within this ecosystem, and being aware of them is a key competence for a project manager.

Understanding Organizational Mindset

The organizational mindset refers to the collective beliefs about intelligence, talent, and capabilities shared by a group within an organization. A group within a company tends to hold similar opinions about intelligence, productive potential, and ability. Certain corporate cultural forms and artifacts specifically attract this type of perspective. These usually include company policies, uniform behavioral norms, power structures, mission statements, and strategies through which the mindset itself is reflected online, be it displayed on a company's own website or a social media platform; Mary C. Murphy provides numerous examples of this.[24]

Murphy's research indicates that an organization's mindset culture consistently impacts five key aspects of how people work together: collaboration, innovation, willingness to take risks and resilience, ethical behavior and integrity, as well as diversity, equity, and inclusion (DEI). These factors, in turn, may impact collaboration efficiency among project stakeholders, which is essential at every project stage and for overall success. This is why it is crucial for project managers to be aware of the culture in which they operate and its level of growth mindset adoption.

Culture, in this sense, is essential as it serves as a fundamental aspect of business that either strengthens or undermines the organization, influencing both external and internal dynamics. A positive, collaborative culture is significant for several reasons: a positive culture draws in talent that aligns with the company's values, it drives engagement and retention, and it affects how employees interact with their work and the organization, thereby influencing their overall happiness and satisfaction.

Cultures of Genius and Cultures of Growth Continuum

As was stated previously, it is vital for project managers to identify the company culture to which they belong. Mary C. Murphy explains that in fixed-mind organizations, or "Cultures of Genius," it is believed that people are stuck in a fixed mindset and incapable of adapting their skills; they either "have it" or they don't. Practices such as "star search" and "stack ranking" are typical results of these fixed-minded cultures. Leadership tends to prioritize acquiring and firing exceptional players at the expense of others. This builds an environment in which workers compete against each other to prove themselves and figure out who the superstars are.

When the term "Cultures of Genius" was used for the first time, most people reacted with excitement to the phrase, especially when it was used in a vague manner. Being a genius is culturally attractive, so it is natural to be drawn to the idea of Cultures of Genius, assuming, often incorrectly, that genius leaders will achieve success. However, Cultures of Genius often produce weaker manifestations of genius. Evidence suggests that people working in these types of organizations are much less able to innovate, be creative, grow continuously, and produce similar results. They tend to foster a culture of perfection. Instead, Cultures of Growth, which are built around complexity, opportunity, and perseverance, may seem more difficult to attain.

Another popular misconception regarding the growth mindset is viewing Cultures of Growth as those in which the leader provides constant support and emphasizes effort over results. Cultures of Growth can be strenuous for those who have reached a learning plateau. The individuals in these cultures would be encouraged to take stock of their growth and transformation rather than merely announcing their success in achieving their aims. They need to see what allowed them to get there, both when they failed and when they succeeded, and use this knowledge to improve the organization—Cultures of Growth offer fundamental, tangible ways to innovate and develop their employees.

These general characteristics of cultures on both ends of the mindset culture continuum were used as points or references for the table below, which describes their possible implications for project management.

Table 3. Organizational Mindset Characteristics and Their Impact on Project Management

Culture of Genius	Implications for Project Management	Culture of Growth	Implications for Project Management
Offers highest performance opportunities.	Focus on project metrics.	Offers highest growth opportunities.	Highly valued: proactive approach to project and self-development.
Emphasizes employees' talent and success.	A strong emphasis on tangible indicators of success, such as titles, awards etc.	Emphasizes employees' motivation and hard work.	High importance of transparent communication regarding project challenges.
Focuses on results.	Important role of producing project documentation and achieving goals.	Focuses on value and processes.	High role of delivering value, process improvements and communication.

Atmosphere of bests: best instincts, best ideas, best people.	Highly valued: competitive approach.	Atmosphere fostering learning, creativity, and resourceful-ness.	Key importance of project knowledge documentation, sharing, and self-development.

Growth cultures can take advantage of the overlap between where people are and where they want to be to leverage this contrast in support of collaborative efforts to move collectively towards shared ends.

Glassdoor—an online platform where employees and former employees can anonymously review companies and their management—analyzed employee data, resulting in companies with a Culture of Genius having lower levels of job satisfaction than those with a Culture of Growth. Murphy's research team also partnered with a management consultancy to conduct interviews with a number of Fortune 1000 companies in various industries, including energy, healthcare, retail, and technology, to assess organizational mindsets. This provided instructive findings: in Cultures of Growth, when employees feel that the culture supports and respects cooperation, they are more trusting and committed. Compared with Cultures of Genius, this culture of interpersonal competition reduces employee loyalty.[25]

Achieving Profit Through a Culture of Growth at Microsoft: A Case Study

Microsoft provides a great illustration of how a growth mindset culture can be a vital part of an organization's success.[26] By the time Satya's Nadella was appointed CEO in 2014, internal silos, bureaucracy, and an ineffective innovation culture afflicted the company. He understood that transforming the culture was the only way to reinvigorate the company's performance. Nadella transformed Microsoft's culture from one of "know-it-all" to one of "learn-it-all," based on Dweck's research.

The cultural shift Nadella implemented at Microsoft was central to the company's renewed commercial success. Under his direction,

Microsoft's stock price rose fivefold, and its market value topped a trillion dollars. This change was tied directly to the growth mindset and to the belief that workers should view setbacks as learning opportunities rather than obstacles. Instead of creating competition, performance appraisals celebrated teamwork and individual achievement, with managers embracing the coaching approach in order to motivate staff.

Experimentation culture grew into a core part of the company's philosophy. Workers were encouraged to test, learn, and be creative, as Microsoft's mission is "to empower every person and every organization on the planet to achieve more."[27]

The growth mindset was also brought to the hiring process, which involved recruiting those who were hungry for learning and innovation. Microsoft's culture encouraged the pursuit of a purpose, with staff not just driven by personal growth but also by the opportunity to address customer needs and be part of the organization's larger cause. With Nadella's focus on inclusivity and teamwork, employees encountered a purpose-driven environment in which they felt a sense of belonging.

The rise of Microsoft's growth mindset culture is a testament to the fact that businesses that value learning, experimentation, and purpose will thrive in this hyper-competitive business world. The company's performance in the wake of this cultural shift can provide other companies with a model that will help them promote an innovation culture that drives their success.[28]

Role of High-Level Processes in Organizational Mindset Transformation

All the dynamics within the mindset ecosystem become even more apparent when an organization undergoes change, which most companies must embrace in order to adapt to new challenges and opportunities. Aaron Sachs, a business analyst, and Anupam Kundu, a strategic financial advisor from the global technology consultancy company ThoughtWorks, identified five critical shifts for

42

organizational transformation that provide a significant context for this process.[29] Their analysis, grounded in Gallup's *State of the Global Workplace* report[30], has become a key reference for studying global trends in enterprise evolution. The shifts described below indicate broader trends that strongly align with a growth mindset philosophy and imply that implementing it will be vital to the future of organizations. Building a growth mindset culture is becoming increasingly essential for companies, leaders, teams, project managers, and employees to thrive.

Shifting from Hierarchies to Networks

In traditional organizations, rigid hierarchies can slow down communication and limit collaboration. With a growth mindset, companies need to implement more flexible, networked ways of working. Networked culture allows collaboration across teams, functions, and locations so that employees can collaborate and share ideas more freely. This evolution is especially crucial in the fast-moving business environment.

For project management, this entails breaking down silos in the team and within the company to create visibility and cross-functionality. The network mindset of the project manager enables collaboration in a team setting and makes the brainpower of the whole company work for them. Consequently, this renders the project team, as well as the organization, nimble and dynamic.

Shifting from Profit to Purpose

Profit is a priority for businesses at all times, but it is not enough in the modern workplace. Employees—especially younger generations—are seeking more than just financial rewards; they want to know that their work aligns with a greater purpose. People are more satisfied when they see their goals aligned with the organization's purpose.

For the growth mindset project team leader, this implies an effort to encourage teammates to look at the big picture; it refers to an interpersonal relationship that inspires people and helps improve team performance. If people have the sense that they are contributing to something more positive, they will be more willing to offer themselves to the team and the company.

Shifting from Controlling to Empowering

In the context of traditional management, managers hold tightly to their decision-making power, which can constrain innovation and ownership. Yet, for an organization to thrive, it must undergo a power shift in which workers are allowed to make decisions and take ownership of their work. Project managers often lead without authority, but they still create their leadership style and can present a more trusting attitude that drives creativity, dynamism, and performance.

Shifting from Planning to Experimentation

Long-term plans need to be frequently updated. The companies that thrive are those that embrace the growth mindset practice of experimenting—not worrying about failure but rather learning from it. For every project team member, the way to achieve this is to stay open-minded, be willing to hear from new voices, and be ready to provide solutions if needed.

Shifting from Privacy to Transparency

Transparency enables open conversations so that employees can see the direction of the organization and how they are struggling with it. This makes them feel closer to the company and more inclined to make a contribution to its success. For a project manager, being transparent means sharing project objectives, issues, and decisions with a project team and stakeholders, as well as involving them in problem-solving and decision-making processes so everyone feels informed, engaged, and respected. Organizational transparency creates a culture of trust and accountability at all levels so that people understand the organization's goals and actively pursue them.

Figure 3. Mindset Shifts for Organizational Transformation

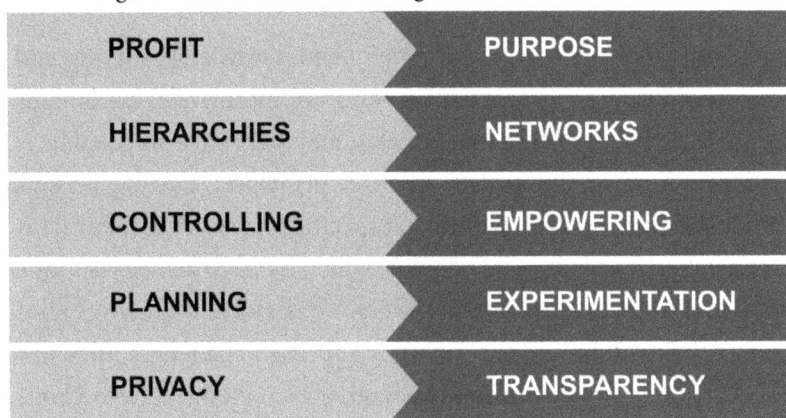

PROFIT	**PURPOSE**
HIERARCHIES	**NETWORKS**
CONTROLLING	**EMPOWERING**
PLANNING	**EXPERIMENTATION**
PRIVACY	**TRANSPARENCY**

Depending on the organization type, project management can either be a central force in the organizational change process or simply act as a recipient of cascading changes and their consequences. Regardless, thinking about these changes in a broader light can make a highly significant difference to the success of a project and to one's career progression. Even if they are not directly engaged in the change, a growth mindset project manager can nonetheless lead mindset shifts within their team and the company by helping people, teams, and the company transition to and welcome change.

Power of Internal Growth Mindset Dynamics

The creation of a growth mindset environment is, in many ways, a project in and of itself. As in every well-executed project, it is good practice to start with an understanding of the situation and context: to specify the scope and identify the constraints. Not every company is willing to adopt a growth mindset, and it usually will not happen overnight. Teams that are used to working within a fixed mindset or who follow highly predictable workflows may struggle with a sudden shift to a completely different approach. In this case, it may be more efficient to start gradually by making incremental adjustments that will serve as a foundation for more significant changes in time.

Team Building for a Growth Mindset Culture

To cultivate a growth mindset-oriented team, knowledge sharing and open communication are essential ways to incentivize continuous learning. It is vital to ensure that all team members have an equal level of understanding of the growth mindset principles and the reasons behind implementing them within that specific team environment. This creates a collective sense of its benefits, which helps to achieve genuine involvement among teammates. If everyone is aware of how mindset can affect their results and the way they approach challenges, they will be willing to commit to it and hold each other accountable. The methods for building this understanding will vary according to the team's needs and preferences. Some teams may prefer instructor-led training or seminars; others may be more interactive, including brainstorming during offsite retreats or "lunch and learn" meetings. Regardless of the delivery method, establishing awareness is key.

When done well, the implementation of growth mindset strategies can not only improve the team's culture but also serve as an entertaining team-building activity. The initial implementation of a growth mindset within a team does not have to occur via a high-visibility, company-wide project; there are many semi-formal opportunities to create a safe space for its introduction. Indeed, every team meeting serves as a way to invite others to bring their ideas into the project, feel ownership over the project's end result, and develop a general growth mindset. Employees are enabled to see themselves as valued and connected and thus to respond to setbacks and missteps with perseverance rather than denial. In addition, making it habitual to review successes and failures during post-project discussions can help turn setbacks into productive lessons. As an effective team practice, if conducted positively and constructively, it can be applied to any project, large or small. This encourages experimentation, lessens the fear of failure, and allows employees to play with solutions—all while creating an atmosphere of openness.

Leading the Way: Inspiring Growth Mindsets in Teams

Developing a growth mindset as part of a project team usually starts with leadership. Leaders of teams need to adopt this mindset in order to establish a culture of empowerment and innovation. They can model attitudes toward resilience and cooperation that will drive their teams to solve problems together and strive for improvement.

Leaders help to encourage a growth mindset by keeping a pulse on team performance, identifying talent needs, encouraging ongoing development, and ensuring the creation of a work environment in which team members can flourish. They maintain close contact with the team to spot and rectify issues quickly, particularly in a demanding environment. Team leaders should be eager to create an environment of openness and curiosity, even if no one is comfortable with being challenged. This culture encourages more efficiency by pushing teams to examine their inefficiencies and test solutions. By communicating clearly and leading effectively, a growth mindset leader can offer a supportive environment in which the team is encouraged to succeed.

Indra Nooyi, who led PepsiCo between 2006 and 2018, transformed the company into one of the largest and most profitable food and beverage businesses in the world. Nooyi was a growth-minded leader with an emphasis on self- and company development, resilience, and creativity throughout her tenure.

The most crucial part of Nooyi's leadership was her focus on performance with purpose in order to enable PepsiCo to grow and make the world a better place.[31] She asked her team to consider long-term objectives, particularly in terms of health and sustainability, and to always keep up with shifting consumer preferences.

Nooyi was also famous for taking setbacks and using them as lessons. When PepsiCo, for example, came under increasing scrutiny for unhealthy beverages and the way their sugary products were harming public health, she guided the business through a profound change. She concentrated on healthier alternatives, expanding PepsiCo's portfolio with healthier products such as Tropicana juices, Quaker Oats, and healthier snacks, as well as downsized products. Her growth mindset was also evident in her approach to managing her team. She urged workers to learn to be leaders and to be looking for opportunities to improve themselves constantly. Even Nooyi was accustomed to receiving constructive criticism and adapting where needed. Despite the difficult times, she always maintained a long-term vision and encouraged her team to innovate and take calculated risks. Nooyi's leadership is a reminder that a growth mindset can be transformative and enable organizations to thrive in the face of immense challenges.

Being a Changemaker

Initiating change within an organization is not confined to formal leadership roles; individuals at all levels have the capacity to drive meaningful transformation. Alex Budak, a faculty member at UC Berkeley's Haas School of Business and the author of the inspirational book *Becoming a Changemaker*, proves that becoming a changemaker is accessible to everyone and begins with a shift in mindset.[32] Change happens by focusing on "micro-leadership"—starting small and taking incremental steps to inspire others. Leadership, in this sense, is not a single act but rather a continuous mindset process. Moreover, a leader is not defined by their position but by their approach, which includes seeing the best in others, building alliances, and seeking advice.

A changemaker mindset, as Budak explains, relates to curiosity, resilience, and optimism. It is rooted in the growth mindset approach introduced by Carol Dweck. This attitude enables individuals to challenge the status quo and turn obstacles into opportunities. Building collaborative connections, focusing on purpose, leading without formal authority based on trust, and psychological safety are critical elements of the journey to becoming an influential leader, regardless of the position.

Project Manager's Day-to-Day Impact

Whether a project manager is a team leader or an individual contributor, they play a crucial role in fostering a growth mindset within the context of a project team. A growth-minded project manager creates a growth-focused team, implements growth mindset practices in tandem with stakeholders, and contributes to an overall organizational Culture of Growth. However, the culture of the organization can determine the extent to which a growth mindset can be integrated into a project. Depending on the organizational situation, the potential to grow can vary, and thus for a project manager it is essential to be aware of these dynamics.

The mindset of the company can affect several aspects of project management: standardized processes, channels for communication, stakeholder dependencies, reporting structures, and job responsibilities. On the other hand, the organizational mindset can be influenced by many factors, including project management practices.[33] These factors, together with growth mindset techniques for shaping a collaborative culture, are listed in the table below.

Table 4. Workplace Culture Factors and Collaborative Techniques of a Growth Mindset Project Manager

Factors Impacting Workplace Culture	Examples	How a Project Manager Can Create a Growth Mindset Culture
Leadership	▪ Communication style ▪ Interaction methods ▪ Leaders' vision ▪ Recognition practices	Identify growth mindset leaders and reinforce this approach within the teams by sharing their inspirational quotes in group chats, showcasing them in company-wide meetings, or posting their conference highlights on social media.

Management Structure	<ul><li>Systems</li><li>Procedures</li><li>Controls in place</li><li>How managers empower and support employees</li></ul>	Managing up by asking a manager for specific support reinforces a growth mindset approach, such as appreciating the effort and not merely the result.
Workplace Practices	<ul><li>Recruitment</li><li>Onboarding</li><li>Compensation</li><li>Rewards</li><li>Training</li><li>Work-life balance practices</li></ul>	Incorporating a growth mindset, building awareness in the training plan for the project team and stakeholders.
Policies and Philosophies	<ul><li>Employment policies regarding workplace</li><li>Dress code</li><li>Scheduling</li><li>Organizational philosophies on hiring</li><li>Performance assessment tools</li></ul>	Voting for including dedication and development in the performance assessment or at least reporting it on a regular basis to build awareness.
People	<ul><li>Personalities</li><li>Beliefs</li><li>Values</li><li>Interactions of employees</li></ul>	Inspiring stakeholders and co-workers to adopt a growth mindset approach.
Mission, Vision, and Values	<ul><li>Communication of the organization's mission, vision, and values</li></ul>	Finding how to link growth mindset principles with the company's mission, vision, and values.
Work Environment	<ul><li>Workplace experience</li><li>Decoration style</li><li>Communal spaces</li></ul>	Ensuring that the physical workplace serves the project needs appropriately and, if needed, suggesting improvements. If possible, using the common areas to communicate and celebrate project progress (not only success).

Communication	• Frequency • Type • Quality	Incorporating growth mindset principles in the project communication style; for example, being focused on the growth aspect when reporting project status, making sure that team members' efforts are acknowledged in the official communication, and constantly improving communication methods to fit the changing situations of stakeholders.

The process of developing an environment that supports growth mindsets is multifaceted and requires an integrated approach and knowledge. The dynamics between organizations, leaders, and teams vary, and a project manager's contribution to a growth mindset can change according to micro and macro conditions, as shown above. There may be limitations, but the essence of a growth mindset is the belief that there is always room to develop it at the individual, group, or organizational level.

Communities Within an Organization: The Role of Employee Resource Groups (ERGs)

Employee Resource Groups (ERGs) are voluntary, employee-driven groups aiming to promote a diverse workplace aligned with the goals of the organizations they serve. These committees create a safe space for employees to share their journeys, discuss struggles, and meet mentors and team members from different business units.

ERGs have been an increasingly important element in promoting a growth mindset within organizations. Originating in the U.S. in the 1960s with Black employees at Xerox organizing around race-based tensions, ERGs have evolved into powerful, employee-led networks that aim to create diverse, inclusive, and supportive work environments.

Today, they are present in 90% of Fortune 500 companies and are critical to all types of personal and professional growth.[34] ERGs typically focus on specific groups of employees based on gender, ethnicity, lifestyle, or personal interests, allowing individuals to bring their authentic selves to work.

Some companies integrate ERGs more deeply into their organizational DNA with distinctive naming conventions. For example, Salesforce has 12 specialized Equality Groups, most of which have a name tied to the company's ethos. "Abilityforce" supports employees with disabilities, while "Asiapacforce" connects employees across the Asia Pacific. "BOLDforce" empowers Black employees, "Earthforce" promotes environmental responsibility, "Faithforce" fosters interfaith understanding, "Indigenousforce" celebrates Indigenous cultures, "Genforce" combats age discrimination, while "Latinoforce" empowers Latinx employees, and "Outforce" advocates for LGBTQ+ equality. Additionally, "Southasiaforce" increases awareness of the global South, and "Vetforce" supports veterans and their families. Salesforce also has the Women's Network and Parents and Families. These ERGs exemplify how organizations can create a sense of belonging.

ERGs develop trust, stimulate open discussion, provoke new thinking about workplace issues, and eventually help manage problems quickly and ease toxic environments. They also create room for career development. The relationships built within ERGs may help recognize and promote talent. To do so, employees and executives must support an ERG, and senior management must serve as sponsors to support and steer the group's actions.

This way, ERGs play a vital role in innovation and organizational success, enriching key processes, such as recruitment, talent management, and leadership development, ensuring a diverse talent pool. By providing growth experiences and leadership development regardless of ethical background, gender, or religious affiliation, companies can continue to develop a culture with a growth mindset.

How Mentoring Programs Elevate a Growth Mindset

Creating a safe space is also a built-in feature of the mentor-mentee relationship. Providing mentoring opportunities has become a corporate standard, and there are proven records of its benefits. Research conducted by Dr. Ruth Gotian and Andy Lopata, as featured in *The Financial Times Guide to Mentoring*, shows that mentoring is highly influential on both individual performance and corporate loyalty.[35] Mentored people make more money, perform better, and are happier in their jobs than non-mentored people. Indeed, 90% of the employees who have been mentored say they're satisfied with their jobs, and more than half (57%) say they're very satisfied. In addition to job satisfaction, 79% of mentored employees feel they are well-paid, and 89% think that their work is appreciated by peers.

Career advancement also seems much more apparent when surrounded by mentors. 70% of those mentored believe their company provides excellent or satisfactory career growth, compared with 47% of those without a mentor. Moreover, mentored employees are less likely to think about quitting.

The influence of mentoring goes beyond personal and professional success. Workforce members who engage in mentoring programs are five times more likely to be promoted within their own company, and mentors themselves are six times more likely to gain promotion. Mentoring also contributes to organizational diversity. Mentoring programs raise minority representation at the management level by 9%-24%, which is better than conventional DEI initiatives that only raise minority representation by 2%-18%. This highlights the impact that mentoring has on career development and the development of a more inclusive and diverse workforce.

Nevertheless, despite what the data says about the great value of mentoring, other metrics say that 24% of the study population don't understand the benefits of having a mentor, and only 37% are ready to become a mentee.[36] The important success factor is choosing the right mentor. Types of mentors and growth mindset opportunities they may offer, described by Dr. Ruth Gotian and Andy Lopata[37], are presented in the table below.

Table 5. Mentor Types and the Growth Mindset Opportunities They Offer

Type of Mentor	Description	Growth Mindset Opportunity
Tormentors	These mentors will give their mentees things to do and keep them busy without working with them or creating a space for cooperation. They can be too caught up in their own tasks to be direct to their mentee and fail often. These individuals can even call out and harass their mentees in the worst cases. They do this in part because they're not confident in themselves; successful mentors measure their success by their protégés' success.	Low
Vanilla Mentors	These are not good or bad mentors; they interact with a lot of mentees but do not really take them to the next level. It encourages stagnation, and people who have had bad mentors are less likely to be motivated to hire a new mentor because there is not much to gain from the current arrangement.	Moderate
Good Mentors	Besides being empathic and responsive to the demands of their mentees, these mentors represent the "Ice Cream" approach: they	Promising

	Introduce, Connect, Engage, Create opportunities, Reply, Encourage, Amplify, and Motivate.	
Great Mentors	Such mentors go out of their way to give their protégés skills, expand horizons, and connect them with others who could be the next step in their lives. The best mentors can predict and be experts at pushing their mentees to go the extra mile.	Exceptional

Mentorship as well as all the approaches listed above promote the intentional seeking out of diverse perspectives, ideas, individuals, and skills other than one's own. While it is hard to eliminate biases at the individual level, adopting a different lens, embracing conflicting views, and debunking presumptions in order to foster cognitive diversity in the organization and eventually opening up within an organization can lessen the effects.

Self-Driven Growth

Although the organization, system, and culture of a company create an environment that nurtures either the development of a growth or a fixed mindset, the mental transitions happen on a micro level—in the realm of our own practices, adaptations, and contributions. These small-scale efforts matter because the consequences of altered individual behavior have a cascading effect and can thereby be magnified.

Self-checking is a way of thinking through critical phases in a project that will help the project manager better understand how to make decisions, take action, and work effectively. There are several tools that can be used to approach this aspect, such as journaling, private feedback sessions, or even simply having quiet time and reviewing triggers and dependencies. Finding an introspection ritual that works best can be helpful in personal and professional development and eventually be beneficial for a whole project team. Growth mindset companies even support the value of self-reflection by motivating employees to document, share, and analyze their thought processes with a manager on a regular basis.

An important part of professional life, which is also a source of insights into areas of potential improvement, is attending self-development events. Taking part in workshops, conferences, or joining networking sessions, including those related to knowledge exchange, helps monitor trends and build skills as well as shows a commitment to professional growth, enabling networking with other people in the field and creating an opportunity to share ideas with individuals from different organizations.

Another growth mindset habit emerges from an eagerness to get more context before making decisions.[38] It is a good practice to take a step back and look around before responding or acting. In everyday work, it is easy to focus on delivering work and forgetting about the bigger picture. Situational awareness is essential to maximizing opportunity. It requires discipline and wisdom to take stock of things, even if we fail to see them. This kind of clarity of thought ensures clear thinking and allows leaders to take risks in uncertain or evolving contexts.

A personal growth mindset also involves acknowledging personal limitations and adapting to changes within the team and organization to drive growth. Project managers often face various pressures, including the expectation to have all the answers. This is where a personal growth mindset is especially valuable, as it allows a project manager to say, "I don't know this yet," and approach challenges as opportunities to learn and grow.

DEI: Establishing Inclusive Workplaces with a Growth Mindset

Adapting a growth mindset in the organization can significantly support diversity, equity, and inclusion (DEI) by promoting an environment that encourages learning, flexibility, and open-mindedness—key elements needed for DEI initiatives to thrive.

According to a 2023 Pew Research Center survey, for a majority of employed U.S. adults (56%), focusing on increasing DEI at work is a good thing.[39] Moreover, more than half of workers (54%) say their company or organization pays about the right amount of attention to increasing DEI. It suggests that the effort U.S. companies have made in the last few years to improve DEI activities was efficient.

What is the role of a growth mindset in supporting DEI processes? The following outlines a few key ways it contributes to fostering a more inclusive, equitable, and diverse environment:

- **Encouraging open dialogue**: A growth mindset develops an environment where everyone can speak freely without being judged. This openness allows project teams to discuss all topics, address biases, and explore new ways to support team members. Leaders with a growth mindset constantly seek diverse perspectives and create psychological safety and a more inclusive atmosphere where everyone feels heard and valued.[40]

- **Promoting resilience and adaptability**: Many projects require adaptation, as teams and individuals may need to confront and challenge long-standing norms or biases. A safe environment makes a space for all team members from all backgrounds to feel empowered to grow professionally without fear of failure or discrimination.

- **Enhancing equity in development opportunities:** A growth mindset unlocks the potential for all team members to learn, regardless of their background. This belief in the ability of every individual to succeed, combined with clear pathways for skill development and mentorship, aligns closely with DEI principles.

- **Reducing bias:** A growth mindset helps to reduce implicit biases in decision-making processes. When project managers focus on potential and improvement rather than fixed traits, it leads to more objective evaluations of team members, which supports fairer hiring, promotions, and resource assignments.

Chapter 4

Resilient Project Management Foundations

A strong foundations in project management are important to set projects up for success. They help structure and ensure that projects are clearly defined, manageable, and capable of delivering the planned results. They also provide a framework to handle the complexity and uncertainty that will inevitably arise during a project's lifecycle.

For the last few years, best practices in the industry have been shifting toward a growth mindset—focusing on flexibility and stressing adaptability. This approach aims to make sure that the project outcomes are closely tied to the bigger picture and that they align with the organization's higher goals and values.

Essential building blocks like managing risks, planning communication, organizing resources, and creating a project plan help keep everything moving, whether the approach is predictive or adaptive. On top of that, data-driven insights allow project managers to make decisions based on accurate data. At its core, a strong project management foundation not only gets projects over the finish line but also supports the organization's long-term growth and keeps everyone aligned regarding what matters.

Industry Standards Evolving with a Growth Mindset Focus

The growth mindset has been widely adopted in the project management world in recent years. One of the largest project management organizations in the U.S. market, the Project Management Institute (PMI), is recognized for its standards, certifications (among them, the famed PMP—Project Management Professional), and tools to help project managers and organizations utilize successful project management practices. PMI creates the Project Management Body of Knowledge (PMBOK® Guide)—a comprehensive yet actionable guide for project managers.

In 2021, PMI released the seventh edition of the PMBOK® Guide, and with that comes a significant philosophical transition to a style of project management that is more principles-based. Where the previous editions focused on a process model with granular knowledge domains, the seventh edition emphasizes 12 core principles and performance areas that mirror different aspects of project management, such as stakeholder and team engagement.

One of the significant shifts is a heavy focus on practices fit for project cases and an increased emphasis on adding value to stakeholders across the entire project lifecycle. The seventh edition also adopts more methodologies such as Agile and Lean, emphasizing inclusivity and adaptability in less technical terms and using more tailored tools and techniques. Respectful project management, collaborative work, and embracing adaptability are all part of the strategies for project success.

Project managers are encouraged to think from the organizational perspective of generating value at the level of the project as a whole. Flexibility and resilience are crucial to adapting to change, and workflow optimization can lead to productivity. If these rules are followed, then project managers will be able to cut through the fuzziness and deliver outstanding outcomes. These updates reflect PMI's intention to align project management with the highly ambiguous environment and to bring a growth mindset into project management discipline.

Value Delivery System: Alignment with Company Goals

Value delivery ensures that the result of a project includes strategic alignment, customer engagement, and a focus on results. Value creation stresses efficiency and resource utilization to achieve the maximum return possible with minimal waste.

Traditionally, project performance was expressed in metrics such as quality, time, and budget or as a ratio among them. While these are all essential considerations in keeping track of a project, as the Project Management Institute states,[41] real success comes from providing the business value that the customer needs. In this light, the key recommendations to deliver value are as follows:

1. Understand the vision.
2. Be clear about the business value of the project.
3. Evangelize the vision and business value to the project team.
4. Foster a team environment to effectively deliver value.
5. Measure the realization of the business value.

Value delivery takes place in the process of information flow, where the strategic directions come from the senior leadership team, and the initiatives are organized into a system of portfolios, programs, and projects. The elements of the value delivery chain create a structured system connecting strategy to execution. Senior leadership establishes the strategic objectives, which portfolios translate into prioritized initiatives, programs coordinate to optimize benefits, and projects deliver specific results—all working in tandem to drive value for the organization.

Following is the overview of the system and the information flow between these elements.[42]

Figure 4. Value Delivery System

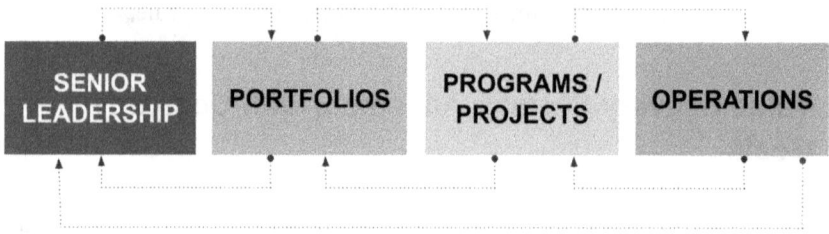

SENIOR LEADERSHIP	PORTFOLIOS	PROGRAMS / PROJECTS	OPERATIONS

Each function in the value delivery system has its specific role:

1. **Senior Leadership** sets the organization's vision, mission, and strategic goals. They prioritize value by investing in programs that support these objectives, providing direction, and taking top-down decisions that influence the organization's direction. They also stimulate building a value delivery culture, fostering accountability, and ensuring that the organization is aligned on metrics.

2. **Portfolio** refers to a collection of projects, programs, and other work that is aligned to achieve strategic objectives. At this level, portfolio managers prioritize and balance resource allocation, monitor risk, and ensure that the collective work aligns with the organization's goals. By focusing on strategic alignment and value potential, portfolio management ensures that each project or program contributes to overarching objectives and maximizes value creation.

3. **Programs** are groups of related projects coordinated to gain advantage and control that could not be done individually. These projects are interconnected and integrated across resources, time, and results to drive the best common value. They drive outcomes to realize benefits relevant to strategy and control project dependencies for streamlined operational performance.

4. **Projects** are the execution layer of the value delivery system, translating strategy into tangible deliverables. Project managers deliver specific outcomes to achieve the defined goals.

Core Project Components

Core project management components are the basic elements that ensure a project is efficiently planned through to completion. These components, such as scope, time, cost, quality, risk, and communication management, serve as a foundation through which virtually every project lifecycle can thus be evaluated. They help the entire project team to stay aligned, organized, and focused on its ultimate goals.

Understanding these components is critical to organizing a project's work and producing quality outcomes. The survey findings from the 2020 PMI *Pulse of the Profession®* report indicate that, on average, 11.4% of investment is lost due to poor project performance.[43] Organizations with mature value-delivery processes are significantly more successful than those without, achieving higher rates of project success. Monday.com, a provider of one of the leading project management software platforms, claims that they are 77% more likely to meet project goals, 67% more likely to stay within budget, and 63% more likely to deliver projects on time.[44] In contrast, those lacking such processes struggle with a higher likelihood of scope creep (47% vs 30%) and project failure (21% vs 11%).

Core project components provide a structured approach to navigating through uncertainties, tracking progress, and dealing with potential limitations in resource availability. By addressing every component, project managers will reduce risks, control costs, and ensure that the project produces quality results on time. Properly managed components mean more efficient processes alongside greatly enhanced customer satisfaction, which results in a much higher likelihood of delivering projects successfully.

To plan effectively, the project's scope should be clear and its objectives should be apparent or set forth in general terms based on the findings or predictions. From there, all those involved should know what outcome everyone wants. A detailed timeline with milestones is necessary, followed by a budget designed along the same lines, maintaining a clear vision for the project. Resources need to be allocated wisely with the correct skills and equipment. Risk management strategies should be developed at an early stage to facilitate the identification of potential problems that may arise, as well as their solutions. Communication channels need to be established for project-wide stakeholder input. Continuous monitoring of these components allows for adjustments and corrections, making sure that the project stays on track and true to its objectives throughout.

The following are the main factors that facilitate effective project execution:

Table 6. Core Project Components

Component	Action for Project Manager to Take
Goals and Scope	Evaluate what the project will include or not include. Objectives, deliverables, and limits are all set here. A clear scope helps to avoid scope creep, where requests for unplanned work made by stakeholders affect the project's deadline, budget, and resources.
Time Management and Schedules	Develop a detailed plan with milestones, deadlines, and task relationships to keep the project on track. This ensures that the project will be finished in time—a deadline is set for your goals.
Expenses and Budgets	Formulate a budget plan to track money and manage expenses. Effective expenditure control allows for making necessary adjustments when unforeseen financial changes happen.
Quality Management	Create procedures and standards of quality for various products to be distributed. This ensures quality is maintained throughout the lifecycle of a project.
Risk Management	Maintain a register of potential risks and plans to deal with them. Active risk management helps to keep nuisances at bay, preventing them from becoming showstoppers.

Communication and Stakeholder Management	Establish a communication plan to keep stakeholders in the loop. Effective communication makes stakeholders enthusiastically support your change.
Resource Management	Plan how to control the resources available, in terms of people, materials, and equipment. Good resource management means that any team member has what they need at the time it is needed, preventing delays.
Plans and Documentation	Additional plans—such as a comprehensive project plan with stipulated milestones, a decision matrix identifying who makes what decisions in an organization or region and how this affects other people or countries, and stages for moving a product to market from research through development and launch—will help bring your team together and give all its members something to strive for.
Metrics and Success Criteria	Define goals through the metrics and the Key Performance Indicators (KPIs) to monitor how well you are doing. These will give any team clear direction.
Governance Structure	Establish clear roles and responsibilities for decision-makers within the project. This governance structure should define who is responsible for oversight, approvals, and escalations. Regular reviews by a steering committee or leadership team will ensure the project stays aligned with the organization's strategic objectives. An established governance framework ensures accountability, transparency, and consistency in decision-making.
Key Business Decisions	Identify the key business decisions that need to be made during the project lifecycle. This includes decisions around resource allocation, major scope changes, prioritization of deliverables, and risk mitigation strategies. Assigning specific individuals or teams to make these decisions and setting up a clear escalation path for high-impact decisions will reduce delays and ensure swift action when required.
Launch Plan	The plan includes pre-launch activities, execution on launch day, and post-launch analysis to optimize the product's market introduction and ongoing performance.

Project Integration Management	Formulate an oversight plan that brings together all key data and merges all project parts in an organized way, which makes one system out of many parts that work together.
	Each element fits together neatly with the others to create a seamless approach that delivers successful projects through an integrated, structured, adaptive project management method.

Navigating Project Uncertainty

In project management, there is always some uncertainty involved that project managers need to deal with, especially in a high-speed environment where results are unpredictable. Uncertainty refers to the gaps in predictions regarding the future, including unknowns about conditions or the impacts of decisions that could influence the project's success. Uncertainty is also associated with ambiguity and volatility, causing unforeseen and rapid changes. PMI defines it as "a lack of understanding and awareness of issues, events, paths to follow or solutions to pursue".[45] This is where a growth mindset can come into play and help project managers and teams succeed in the face of the uncertainty that comes with complicated projects.

Uncertainty is caused by different factors, including market instability, political influences, lack of information, shifting stakeholder expectations, technological or regulatory change, gaps in product functionality, or changing availability of resources. The Covid-19 pandemic proved that uncertainty might have an unpredictable scale, impacting the ecosystem of business organizations, starting with the availability of resources.

The impact of uncertainty can emerge in the form of unplanned costs, time delays, change of scope, lack of resources, or anything else that can throw a wrench into a project. In contrast to risks, which are typically recognizable and quantifiable, uncertainty is more diffuse and tends to involve unknowns that can't be accurately forecasted or prepared for. Thus, project managers have to devise ways of dealing with these unknowns and using experience, intuition, and agility to make sound decisions within an evolving landscape. Both risk management and uncertainty management are critical skills for all project managers to master.

Uncertainty vs Risk

"Risk" refers to the "uncertain event or condition that, if it occurs, has a positive or negative effect on one or more project objectives."[46] Risks are generally measurable, quantifiable, and predictable. They may have different noticeable effects, ranging from the project to the program and organizational levels; risk management is about identifying and prioritizing these risks.[47] Subsequently, appropriate steps are taken to reduce or leverage the impact of the risks. Essential project management risk management strategies are intended to identify, measure, and react to risks to minimize their negative impact on projects. Such risk mitigations and opportunity-recognition strategies drive project success. The table below includes the most common risk response strategies.

Table 7. Risk Mitigation Strategies

Mitigation Strategy	Description	Example Scenario
Avoid	Modify project plans or change the project approach to completely eliminate the risk.	A project team decides to avoid risk using a new technology to prevent technical failure.
Escalate	Transfer the risk to a higher authority or stakeholder that is better positioned to handle it.	A project encounters a regulatory risk that needs executive team intervention.
Transfer	Shift the risk to a third party, such as a vendor, supplier, or insurance provider.	An organization purchases insurance to cover potential financial losses from project delays.

Mitigate	Take proactive actions to reduce the probability or impact of the risk.	A project team conducts additional training to reduce the risk of human errors.
Accept	Acknowledge the risk and prepare to handle its impacts if it occurs, usually when the risk impact is minimal or unavoidable.	A project team accepts minor delays due to potential supply chain issues, as they have minimal impact on the timeline.

"Uncertainty" is the broader term that refers to the condition in which a project manager has no knowledge of or understanding about all that will occur in the future and is incapable of estimating or modeling all possible outcomes. The dynamic nature of uncertainty requires project managers to quickly assess and understand the implications of the current situation. David Hillson, an author of the *Managing Risk in Projects* book, notices that not all uncertainty is risk; all risks are uncertainty.[48] While risk management focuses on "known unknowns," uncertainty requires an open-ended approach that accepts the unknown and the unpredictable.

Uncertainty arises from the future, which is never entirely predictable, because there is not enough data or information is ambiguous. The management of uncertainty needs to be handled strategically, with flexibility, agility, and taking into account the best available information that is continually updated as information comes in.

Researchers from the INSEAD Technology Management Institute in Singapore—Arnoud De Meyer, Michael T. Pich, and Christoph H. Loch—conducted a study on uncertainty and identified several distinct types of uncertainty that organizations face. Their research highlights how different forms of uncertainty can impact project decision-making and overall company performance:[49]

- **Variation**: Small influences cause variations in activity durations (e.g., delays due to weather or illness). The plan is detailed, but schedules and budgets can shift.

- **Foreseen uncertainty**: Identifiable but uncertain factors (e.g., drug side effects). Managers prepare contingency plans but may not use them if the risk does not materialize.
- **Unforeseen uncertainty**: Unexpected events or interactions that cannot be anticipated (e.g., Viagra's market shift from heart medication to sexual performance drug).
- **Chaos**: Fundamental uncertainty at the project's core, where the goals and plan may evolve as the project progresses (e.g., Java's shift from appliance control to a web programming language).

Anticipating Uncertainty in Projects

When managing uncertainty, project managers need to implement measures that make it as low a risk as possible to project delivery. The growth mindset is vital here: in such situations, it will foster a strong team dynamic and enable workers and project managers to stay flexible when faced with setbacks. Collaboration helps refine tactics and bring them into line. At the same time, a problem-solving attitude keeps the team focused on practical issues.

A spectacular example of navigating uncertainty is the story of Amazon. As an agile company, Amazon has thrived on a belief in uncertainty and changing its business models accordingly. By the time the company went from an online bookshop to an e-commerce platform, it was still operating in a very uncertain market. Amazon adopted flexible project management processes, like prototyping and iterative development, which allowed them to evolve their approach as the market changed. Because they prioritized predicting the risk in such fields as supply chain management and technological infrastructure, they were able to maintain growth in volatile markets.

The following are some selected techniques that may help to navigate uncertainty.

Table 8. Uncertainty Managing Strategies

Technique	Description	Example Scenario
Scenario Planning	Developing multiple future scenarios and planning responses for each, preparing teams for different outcomes.	Creating different project timelines based on optimistic, realistic, and pessimistic forecasts of progress.
Stakeholder Engagement	Regularly communicating with stakeholders to align expectations, gather insights, and adjust direction as new information arises.	Scheduling weekly check-ins with stakeholders to ensure project goals align with evolving business requirements.
Diversification	Spreading risks across different areas or aspects of the project to reduce the impact of uncertainty on any single element.	Allocating resources to multiple vendors or markets to avoid over-reliance on one supplier or customer base.
Leverage Agile Approach	Adopting Agile principles to break the project into small, iterative cycles, enabling quicker adaptation to change and uncertainty.	Using iterative cycles to develop software, with regular feedback and adjustments after each Sprint to align with changing requirements.
Contingency Planning	Preparing backup plans and allocating resources to manage unforeseen challenges without derailing the project.	Developing a contingency plan for unexpected supplier delays, including alternate vendors and expedited shipping.
Flexible Resource Allocation	Ensuring resources can be easily adjusted or reallocated to accommodate changing project needs.	Reassigning team members from less critical tasks to areas facing delays to maintain project timelines.

Stress Tests	Evaluating the project under extreme conditions to identify vulnerabilities and prepare for worst-case scenarios.	Running simulations to test how a new software application performs under high user traffic to ensure system stability.
Project Environment Monitoring	Continuously observing external factors and project conditions to identify emerging risks or changes that may require adjustments.	Tracking market trends, regulatory changes, or competitor actions that could affect the project's success.
Adapting Growth Mindset Principles	Fostering a culture of continuous learning, resilience, and adaptability, where challenges are viewed as opportunities for growth.	Encouraging the team to accept setbacks as learning experiences, adapt to change quickly, and collaborate effectively to solve problems.

A growth mindset enables project managers to see uncertainty as an opportunity to drive innovation, flexibility, and iteration. This mindset makes projects easier to manage, and it makes us more self-reliant in the face of future challenges, both personal and professional.

Measurement and Performance: Using Data for Growth

Understanding how to configure strategic and tactical measurements is key to project performance and organizational success. "What gets measured, gets managed," said Peter Drucker, the guru of management theory, which was why measurements were deemed essential in planning projects. When done well, this enables teams to dig deep into the important aspects and make it work. KPIs provide directionality, but the path to achieving it depends on the specific goals defined in project management. This is not only a basic value—it is also sometimes referred to as a "golden rule of project management" because it makes the difference between surviving and thriving.

Clearly defined goals set expectations and direct project efforts, but the SMART technique provides the structure. These criteria—Specific, Measurable, Achievable, Relevant, and Time-bound—are the backbone of effective metrics. They provide key insights into project health, allowing teams to make data-driven decisions. As described in an Adobe Workfront whitepaper, "Without effective methods for prioritizing work, you're speeding down a highway without headlights."[50] This is undoubtedly correct—metrics and KPIs help the team focus on the top priority tasks, leading to the achievement of strategic project results.

Goals are critical in today's business environment. Companies invest in data-driven decision-making frameworks in which goals are pivotal. A study from 2005, quoted by the *Harvard Business Review*, indicated that approximately 60% of organizations needed more explicit financial goals aligned with their strategic priorities.[51] In contrast, a 2015 survey conducted by Donald Sull and similarly published by the *Harvard Business Review* that involved nearly 8,000 managers across more than 250 companies revealed that over 80% of managers reported having clear, measurable goals and sufficient funding to achieve them.[52] This shift highlights the increasing acknowledgment of the significance of clearly defined goals in propelling organizational success.

Goal-setting theory, originally developed by Edwin Locke and Gary Latham, emphasizes that specific and challenging goals lead to higher performance than vague or nonexistent ones.[53] The fundamental principles of the theory include clarity (goals should be clear and specific), challenge (goals need to be challenging), commitment (individuals must be committed to achieving the goals), feedback (regular feedback on progress), and task complexity (goals are to be split into manageable parts).

Establishing success measures and KPIs facilitates effective monitoring and fosters a growth mindset. Regularly reviewing metrics empowers a culture of continuous improvement, helping teams adapt and enhance processes. Moreover, metrics should align with broader organizational goals to ensure project outcomes contribute to overall success. They also serve as a common language for communicating progress to stakeholders and help identify potential risks early on, enabling proactive mitigation. Benchmarking against industry standards or past projects provides valuable context and highlights areas for improvement.

Key Focus Areas for Choosing Appropriate Metrics

Choosing the right project metrics is crucial for effectively measuring success in a business environment. It is not just about tracking progress; it is also about empowering project managers to make data-based decisions and navigate the project towards success.

Table 9. Types of Project Metrics

Metric	Description
Financial Metrics	Measures used to assess a company's financial health, performance, and profitability, evaluating the organization's overall effectiveness. These metrics help stakeholders make informed decisions regarding investments, budgeting, and strategic planning. Examples: Annual Contract Value (ACV), Revenue Growth Rate, Annual Recurring Revenue (ARR), Gross Monthly Recurring Revenue (GNMRR), Profit Margin, Earnings Before Interest Taxes (EBIT), etc.
Product Metrics	Focusing on the product itself, assessing aspects such as development timelines, feature utilization, and product quality. This helps ensure that the project meets the intended product specifications and user expectations. Examples: Daily Active Users (DAUs), User Retention Rate, Churn Rates, Feature Adoption, etc.

Customer Metrics	Focusing on understanding customer satisfaction and engagement, as these can provide valuable insights into how well the project serves its target audience. Examples: Customer Satisfaction (CSAT), Net Promoter Score (NPS), Customer Retention Rate, Customer Churn Rate, etc.
Project Performance Metrics	Directly assess the project's progress and health, including adherence to timelines, budget compliance, and resource allocation. This is essential for project managers to ensure that the project is on track. Examples: Schedule Variance (SV), Cost Variance (CV), Planned Value (PV), Actual Cost (AC), Return on Investment (ROI), Resource Utilization, etc.
Sales Metrics	Sales metrics show how a project affects revenue, helping businesses assess the success of their sales strategies. Examples: Total Sales Revenue, Sales Growth, Conversion Rate, Sales Cycle Length, Win Rate, Churn Rate, Pipeline Value, Sales Qualified Leads, etc.
Marketing Metrics	Marketing metrics help measure the success of marketing campaigns and evaluate overall effectiveness of marketing activities. Examples: Number of Leads Generated, Cost per Lead (CPL), Marketing Qualified Leads (MQL), Conversion Rate, Website Traffic, Bounce Rate, Email Open Rate, Click-Through Rate (CTR), Cost per Click (CPC), Organic Traffic, etc.
HR (Employee) Metrics	Employee metrics aim to understand team performance and engagement, providing insights into the team's dynamics and morale. Examples: Productivity, Employee Satisfaction, Turnover Rate, Attendance, Collaboration, etc.

Organizations can track progress and make smart decisions that fit their goals by choosing and using the right metrics. The list in the table above is not complete; other metrics might be necessary based on the specific project and the team's goals.

Leveraging a Growth Mindset for Project Metrics

Metrics are not just numbers; they are tools for growth. To embrace a growth mindset, one can leverage metrics to learn, adapt, and succeed. Using a growth mindset to set project goals is also about embracing new ideas and taking risks. As outlined in the second chapter of this book, viewing challenges as opportunities encourages new ideas and experimentation. This process allows teams to enhance their products based on user feedback, leading to better decisions in the future.

A growth mindset emphasizes the importance of team development and collaboration, enabling members to tackle new challenges and share insights across functions. The collaborative approach creates an environment for team members to stay empowered and utilize their problem-solving capabilities. By investing in team development, organizations encourage ongoing learning and skill enhancement, which prepares employees to handle challenges more effectively. Sharing knowledge becomes a part of the workflow in a collaborative organizational culture. This mindset fosters resilience, curiosity, and long-term growth, making it essential for effective project goal-setting.

Using a growth mindset to measure project outcomes includes the following:

- **Adaptability**: Viewing scope/direction changes as learning curves and adapting metrics to new objectives.
- **Learning from mistakes**: Driving improvement with data-driven insights to uncover issues and rectify them, thereby improving the process over time.
- **Creating innovation:** Encouraging experimentation and building metrics to get even better performance information.

- **Accountability**: Enabling the ownership of measures to enhance participation and alignment with project outcomes.
- **Open collaboration**: Encouraging open feedback and collaboration for more meaningful and precise metrics.
- **Strategic alignment**: Achieving long-term organizational objectives by prioritizing metrics that drive results.
- **Resilience**: Using failures as a reason to research and optimize performance.

By reviewing metrics regularly, a project manager can adjust their activities to achieve strategic organization goals.

Chapter 5

Building a Dynamic Project Management Framework

A dapting a suitable project management framework is essential, as it provides a systematic method for successfully planning, managing, and delivering projects. A well-designed framework helps ensure good execution, the effective deployment of resources, and alignment with business objectives. Project managers, depending on the organizational context and project goals, can choose between predictive and adaptive approaches or, as has become popular in an uncertain environment, blend these two and make a uniquely tailored framework.

The PMI *2024 Pulse of the Profession* report points to a paradigm shift in project management in which fit-for-purpose development has taken the lead.[54] Though more complicated, it rewards those companies that see the opportunities to improve their project management. The report also emphasizes the importance of mixing different methodologies to maximize project management strengths, such as predictability, adaptability, and innovation; this is of high importance as digital transformation accelerates in multiple industries. With the rise of hybrid models in many companies, people know what works and how to leverage different tools and methodologies to ensure consistent project results. However, the report concludes that there is a diminishing return on investment if only one strategy is adopted. Also, standardized processes, including risk management and stakeholder engagement, are still crucial for success. Businesses with these practices consistently show better-than-average project results, each employing the specific blend of agile and predictive methods that fit their business context.

Agile is the leading framework for adaptive project management, while the Waterfall framework is the leader for predictive strategies. Both methodologies remain widely used in project management, with comparable numbers of organizations employing each.

Predictive Approach: Waterfall

Predictive project management is an approach based on a linear process. It is designed to operate with projects that have a precise sequence, where each step continues from the previous. It is rooted in a planning and documentation-driven management system, with a blueprint to lead each step of the process from beginning to end. This rigidity allows predictable results, which are defined at the early stage of the project. The predictive approach is analogous to house construction: everything is planned before construction even begins, and anything that needs to be changed in the middle of the process would create extra delays and costs.

Predictive project management can be used in all types of organizations. Regulated sectors favor this model for its predictability in projects demanding accuracy and oversight. Construction giants Bechtel and Fluor rely upon their emphasis on careful frontloading for massive infrastructure efforts that require exhaustive pre-planning. In the same way, aerospace leaders such as Lockheed Martin and Boeing employ their disciplined stages to meet the industry's stringent standards and testing requirements for aircraft and weapons, as well as to develop safety-critical technologies.

The automotive industry leans on the predictive approach's sequential nature for vehicle architecture and manufacturing streams, in which regulated uniformity proves pivotal, as automakers Ford and Toyota have found. Pharmaceutical heavyweights Pfizer and Johnson&Johnson

also opt for the use of controls in regulated drug making, from research through trials and approvals, tailoring development to meet guidelines at each stage. Government missions routinely employ it for documentation-intensive undertakings with fixed scopes managed by agencies like NASA and the U.S. Department of Defense. Telecom carriers like AT&T and Verizon sometimes tap the approach's virtues when network builds involve prolonged design and rollout that call for meticulous project management. These players select the predictive methodology when compliance is needed, and knowing all the project phases and the outcomes upfront is preferred.

Even in the tech industry, while many companies favor an Agile approach to project management, certain endeavors remain well-suited to the traditional predictive methodology or blended models. Predictive project management delivers advantages for initiatives involving product development with static needs through comprehensive pre-planning and well-defined schedules, partitioned into discrete stages from requirement-setting to testing and deployment. Such regimented management assures a judicious allocation of assets and fulfillment of goals in compliance with quality expectations.

Phases of the End-to-End Predictive Projects

The use of five project management phases provides a structured means of getting projects completed successfully. Projects, in the predictive model, start with the Initiation phase, in which a project charter helps to assess feasibility and business value, as well as identify the right stakeholders. Following this is Planning, which involves creating detailed plans. This means setting objectives, defining tasks, and developing schedules. In the next phase, Execution, the project plan comes alive as resources are allocated, teams are organized, and deliverables are created, with the project manager making sure that work corresponds to objectives as scheduled. Monitoring and Controlling go alongside Execution. Measuring performance also involves addressing issues and making decisions on any slight adjustments needed to keep the project rolling along smoothly. The final phase, Closing, formally closes the project with the handover of deliverables, the release of resources, the collection of documents, and a gathering at the post-project review for learning for next time.

Figure 5. Five Phases of Predictive Projects

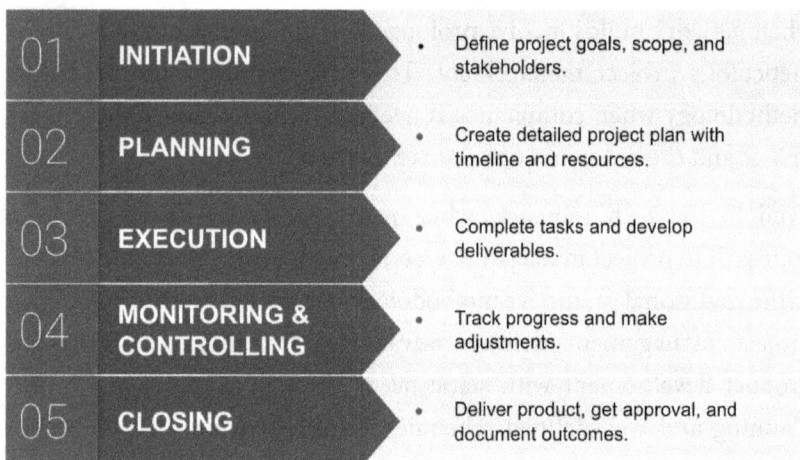

01	INITIATION	• Define project goals, scope, and stakeholders.
02	PLANNING	• Create detailed project plan with timeline and resources.
03	EXECUTION	• Complete tasks and develop deliverables.
04	MONITORING & CONTROLLING	• Track progress and make adjustments.
05	CLOSING	• Deliver product, get approval, and document outcomes.

Waterfall Framework

The Waterfall model is the main framework of predictive project management. This approach splits the project into various steps— **Requirements Gathering > Design > Implementation > Testing > Maintenance**—where each phase must be finished before the next one begins.

The Waterfall method requires detailed documentation and planning at every stage so that project managers can predict failure and lay out expectations early. This framework, similar to all predictive project management methods, is used for projects that have clear requirements and low levels of uncertainty and, thus, those in which the life cycle of a project is known. After the phase is complete, changes are difficult to make. Project managers need to establish a complex and formal change control mechanism to manage changes. Each change requires an impact analysis to review how it impacts the other project areas: time, budget, scope, and resources. The Waterfall technique consists of the following fundamentals:

- Streamlined schedule: Each step needs to be successfully completed before the next.
- Little room for error: Limited change flexibility after the phase is completed.
- Roles are clearly defined: Every team member has a clear role.
- Prioritization of project planning: The Waterfall approach focuses on preliminary project planning.

The main advantage of the Waterfall model is that it is centered around documenting everything, which creates a well-defined process for project execution and makes it easy for stakeholders to share information. The Waterfall model's systematic structure also increases accountability by creating separate deliverables and milestones for each step, letting project managers to track and evaluate easily. Such predictability allows them to focus on execution and stick to timelines without the fear of scope creep. Simply put, when used correctly, the Waterfall model is an excellent platform for plan-based, cost-effective project delivery with visibility at all stages of the project lifecycle.

Navigating Waterfall Stages: Example

Let's consider the case of a software development project to present the subsequent phases of the Waterfall. The project will begin with the **Requirements phase**, in which project managers, business analysts, and stakeholders set out the requirements to be collected and clarified. This is the most important step in creating an exact and complete project scope. It defines requirements, constraints, and project goals. This phase results in a final requirements document, which becomes a project roadmap and is used to get everyone aligned on what the project involves.

Next comes the **Design phase**, which turns the requirements into an actual project blueprint. This step is carried out by system architects and designers, who develop high-level and complex designs of the system architecture, UI (User Interface), data structures, and communication among components. This planning is very important so as not to make any missteps and avoid rework in the future. The design documentation that is created in this step informs designers in the coding step below.

In the subsequent **Build phase**, developers translate design specifications into actual code. This is when code is written, components of the system are assembled, and unit tests are conducted to ensure all modules are working properly. This phase focuses on the execution of designs into executable software, and it takes a dedicated group to keep things in sync with the documented specs.

When the software is completed, it goes into the **Test phase**, and everything is tested thoroughly to ensure it's working and stable. It may include different types of testing, such as unit testing, system testing, integration testing, and UAT (User Acceptance Testing). Testing is to detect the issues, verify deliverables against requirements, and determine whether the code can be deployed.

After testing is complete, the project goes into the **Implement/Deploy phase**. The code gets put in production. This is where the rollout is fully planned, and users are trained, and all of this is managed to ensure that a seamless migration to the new system is minimally disruptive to the users. This can be a full rollout or a gradual implementation based on the scope and number of users of the project.

The last phase is **Maintain**, in which the team ensures the long-term viability and functionality of the product after deployment. This stage is followed by continued support, regular updates, bug fixes, and improvements in accordance with users' needs and evolving requirements.

The following figure presents the sequential phases of the Waterfall software development project.

Figure 6. Waterfall Project Framework

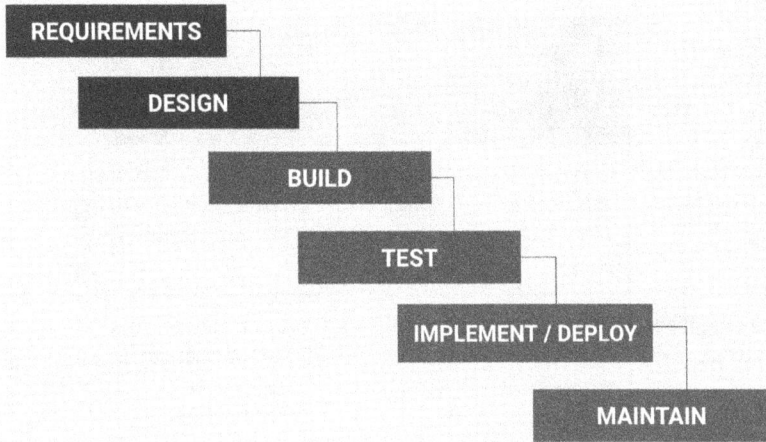

Applying the Growth Mindset Mentality in Predictive Project Management

When predictive project management incorporates the growth mindset principles, it may lead to both project success and engagement from the team. With a culture that supports learning and continuous improvement, team members are empowered to contribute and innovate throughout the lifecycle of a project.

In predictive project management, where scheduling and deadlines are critical, a growth mindset allows team members to perceive setbacks as learning opportunities rather than as obstacles. This attitude fosters transparency, creates a culture of feedback, and is an excellent way to solve problems. A growth mindset project manager shows resilience and flexibility and encourages their teams to welcome change and continue to improve. Ultimately, implementing a growth mindset into predictive project management improves individual and team performance, project delivery, and overall organizational improvement. The table below shows examples of incorporating the golden rules of growth mindset project management into Waterfall projects.

Table 10. Applying Golden Rules of Growth Mindset Project Management to Waterfall Projects

Growth Mindset Golden Rule	How to Add to Waterfall Projects	Example
Continuous Improvement	Conduct "lessons learned" sessions after each phase and apply findings to subsequent phases.	After the design phase, refine templates to improve planning for future projects.
Learning from Feedback	Collect stakeholder feedback at milestones and adjust processes or outputs accordingly.	Use feedback from the testing phase to improve deployment documentation.
Creating a Collaborative Culture	Promote cross-functional collaboration during key phases like requirements gathering.	Host workshops to align business and technical teams early in the project.
Fostering Innovation	Encourage brainstorming sessions before locking the deliverables for each phase.	Conduct ideation meetings in the planning phase to explore creative solutions.
Taking Risks Without Fearing Failure	Allow buffer time for experimentation within structured phases.	Test alternative solutions during prototyping with the freedom to fail.
Thriving in Ambiguity	Use scenario planning in the initiation phase to prepare for uncertainties.	Develop contingency plans for high-risk project areas.
Embracing Challenges as Opportunities	Frame challenges as improvement areas during phase reviews.	Address resource constraints during execution as opportunities to optimize processes.
Bouncing Back from Setbacks	Maintain and regularly update a risk register with mitigation strategies.	Realign timelines and adjust dependencies to recover from delays.

Facilitating Knowledge Transfer	Create a shared repository for project documentation and lessons learned.	Use platforms like Confluence to centralize project insights for team access.
Leading by Example	Demonstrate adaptability and openness to learning throughout the project lifecycle.	Share personal learning experiences to inspire the team to cultivate growth-oriented behaviors.

A Growth mindset can also be blended into predictive project management by leveraging growth mindset tools and methods such as Agile, Lean, or Kaizen. These approaches are outlined in the following chapters of this book and offer a roadmap for creating an agile, resilient, and adaptive culture in predictive project teams.

Adaptive Approach: Agile

Adaptive project management focuses on adaptability and ongoing learning, which enable project teams to alter their plans and tactics dynamically as project requirements and external conditions shift. Agile is one of the most common adaptive methodologies, emphasizing iterative development, teamwork, and frequent feedback, all of which make projects successful.

The Agile Framework

Agile project management is an iterative delivery process focusing on constant releases and customer feedback. Its flexibility to make changes at each iteration increases its speed and flexibility compared to the more traditional Waterfall project management methodology, which is rigid and linear. With the needs of today's customers and businesses to react rapidly, agile project management allows for the ability to change and adjust processes throughout development. Due to the popularity of Agile project management, more organizations are adopting it for single teams or projects, as well as across entire programs. Agile has even been extended beyond development teams to IT, marketing, business development, and other sectors.[55]

The advantages of Agile project management make a large difference in the project's success:

- It allows for faster feedback cycles and helps teams identify problems earlier, increasing customer satisfaction.
- Agile helps improve time to market so that products and services can be released quickly.
- The approach significantly supports accountability and transparency among team members so they feel heard and held accountable for their work. In addition, project teams within Agile processes boost productivity.
- Agile's flexibility in prioritizing value delivery ensures the priority remains on core features and work items, helping project deliverables align with customer needs.

The *State of Agile Report 2023* proves that among teams that adopted the Agile framework, seven in ten report satisfaction with Agile practices at their company, citing increased collaboration as a key factor, while over half attribute their satisfaction to better alignment with business needs.[56]

Origins of Agile Thinking

Agile project management originated in the early 2000s when a group of software engineers sought to reform the broken project management paradigm. In 2001, they created the Agile Manifesto, which laid out the values and principles of Agile. It has four fundamental values:[57]

- **Individuals and interactions over processes and tools**. Agile focuses on people and communication rather than process and tooling. Methods and tools are essential, but collaboration is the key to software development success.
- **Working software over comprehensive documentation.** Agile is about providing actual software for real customers,

not extended documentation. Agile is fond of documentation, but it should be there to assist, not prevent it.

- **Customer collaboration over contract negotiation.** Agile requires continuous engagement with customers through the development process. However, instead of focusing on detailed specifications, Agile teams partner with customers to iterate and correct the project.
- **Responding to change over following a plan.** Agile is flexible. Teams understand that customers and the market can change throughout the project lifecycle. Flexibility is valued over strictness in following a plan in the beginning.

The Agile Manifesto also includes twelve principles. Its creators highlight the need to prioritize customer satisfaction through continuous software delivery, embracing changing requirements, and delivering smaller outcomes frequently rather than all outcomes simultaneously. It emphasizes motivated teams built without formal structures and face-to-face communication and promotes simplicity and regular improvements.

Key Agile Phases

The Agile method consists of several iterative steps: **Plan > Design, Develop >Test > Review > Deploy**. During the Plan phase, teams set objectives, identify requirements, and prioritize work to keep it aligned with the project vision. The Design stage follows, where ideas are generated, and prototypes or wireframes are drawn. The Develop phase is where the code development happens, and teammates collaborate to implement the design. In the Test phase, the quality and performance of the deliverables are tested. Next comes the Review phase, which gives feedback and suggestions for changes. Finally, the Deploy stage launches the product out into the world, ending the iterative process and allowing the team to take steps to refine based on user feedback. The following diagram presents key Agile phases.

Figure 7. Agile Development Model

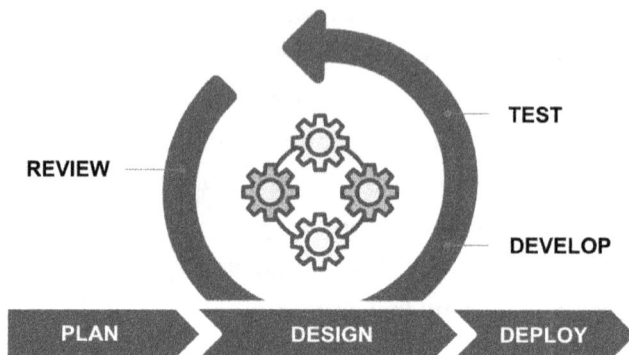

Scaled Agile Framework

Agile can be scaled for more matrixed organizations. SAFe—Scaled Agile Framework—is designed to help large teams implement Agile practices across multiple departments. SAFe provides a structured approach that ensures alignment and collaboration at various levels. Key elements of SAFe include Agile Release Trains (ARTs), which synchronize multiple teams working toward a common goal, and Program Increments (PIs), which break down work into timeboxed planning and delivery cycles. The framework also integrates Lean Portfolio Management (LPM) to help prioritize and fund the most important initiatives, as well as continuous delivery pipelines to support faster delivery. By adopting SAFe, organizations can scale Agile practices, making their integrated processes more flexible and adaptive.

Integration of Agile and Growth Mindset

Incorporating the Agile methodology into project management frameworks helps build a growth mindset-oriented organization. A growth mindset encourages individuals and teams to think of obstacles as a learning experience. This is in tune with Agile's emphasis on agility, continuous improvement, and change management. Growth-mindset teams are more open to being coached, experimented upon, and learned from, thus producing innovation. Teams can improve collaboration, creativity, and productivity by creating an environment in which individuals can take risks and leverage the lessons learned in the process.

One of the critical aspects of the Agile methodology is the concept of self-organizing teams, where members collaborate, take responsibility, and respond quickly to changing demands without relying on top-down management. In these teams, members share their efforts, take responsibility, and respond quickly to changing demands. Each person brings something different, and collectively, the team decides what to solve and how to handle problems. That freedom creates more flexibility and a sense of accountability, enabling faster decisions. Focused on communication, transparency, and trust, autonomous teams utilize Agile to thrive and foster high engagement and innovation.

Agile vs Waterfall

The Waterfall and Agile approaches are two project methodologies suitable for different purposes. More than 65% of organizations utilize Agile, while over 62% rely on Waterfall approaches, according to a study conducted by PM Solutions Research in 2022.[58] This indicates that both methodologies continue to play a significant role in organizational practices, with many organizations using these two approaches at the same time, adopting a hybrid model or selecting the methodology that best aligns with the specific requirements of their individual projects.

The Waterfall method has a linear, sequential structure with fixed phases, such as requirements, design, and testing. It depends on extensive upfront planning and documentation. This approach limits customer involvement after the initial phase. Changes are difficult to implement and can lead to longer delivery times. In contrast, the Agile approach takes an iterative approach based on flexibility and continuous re-planning, whereby stakeholders are deeply involved. Regular feedback allows teams to accommodate changing requirements and produce working features in increments rapidly. While the Waterfall approach is suitable for projects with stable requirements, the

Agile excels in dynamic environments. The table below illustrates the fundamental differences between the Waterfall and Agile project management methodologies.

Table 11. Agile vs Waterfall Approach Comparison

Aspect	Agile Project Management	Waterfall Project Management
Structure	Iterative and incremental.	Linear and sequential.
Project Phases	Cycles (Sprints) for development and feedback.	Fixed phases (requirements, design, implementation, testing, maintenance).
Flexibility	Highly flexible; changes can be incorporated throughout the project.	Fixed; changes are difficult to implement once phases are complete.
Planning	Continuous planning throughout the project lifecycle.	Comprehensive upfront planning.
Customer Involvement	High involvement through regular feedback and collaboration.	Limited involvement after the initial requirements phase.
Risk Management	Continuous risk assessment and adaptation throughout the project.	Risks identified during initial planning; challenging to adapt to new risks.
Documentation	Minimal documentation; focus on frequent deliverables.	Extensive documentation is required at each phase.
Delivery	Incremental delivery of product features throughout the project.	The final product delivered at the end of the project.
Team Dynamics	Self-organizing teams; no strict hierarchy.	Hierarchical structure; roles are well-defined.

Time to Market	Faster; regular delivery of features allows for quicker market entry.	Longer; delays can occur if requirements change.
Quality Assurance	Continuous testing and integration throughout the development process.	Testing occurs at the end of the development cycle.
Customer Feedback	Limited feedback after the initial requirements phase.	Regular feedback after each iteration or Sprint.

Tailoring: Hybrid Models

Project managers can leverage the tailored, hybrid approach to improve project management effectiveness, efficiency, and utility by adapting the frameworks to the nature and requirements of each project. Since each project is different in size, complexity, risk, industry, and stakeholder requirements, a generic approach will result in inefficiencies, waste of resources, or lack of oversight.

The PMI organization defines tailoring as "the process of referencing framework documents, standards, and other relevant sources and utilizing those elements that provide processes, tools and techniques that are suitable for that particular organization."[59] According to PMI research, projects that were managed without a well-defined project management methodology achieved project success rates of 66%. In contrast, projects that were managed with a specified project management methodology achieved project success rates averaging 74%. Organizations that utilized a fully tailored or customized methodology reported a project success rate of 82%.[60]

Tailoring involves the adjustment of current processes that are presently utilized by the organization. Therefore, tailoring is the act of customizing a project management methodology. Instead of putting a pre-determined model into practice, tailoring allows project managers to select elements from different frameworks that suit each project's needs, considering scope, complexity, stakeholder needs, and resources. This involves evaluating the project landscape, selecting the best processes to use, and determining which practices are appropriate to reduce or eliminate. For example, a small-scale Agile project may only require basic documentation and simple, fast communication, while a larger project may require a robust governance model and granular tracking mechanisms.

In some companies, project management strategies vary depending on the nature of the project. For instance, at Google, project management strategies change depending on the nature of the project. As an example, the Agile process is used for software development efforts, including the Google Cloud and Android teams. These kinds of projects need to be characterized by fast iteration and adaptability, which Agile (Scrum, Kanban, etc.) is best suited for. By contrast, larger hardware projects, such as Google's Pixel phone or Nest thermostats, adopt more predictive approaches, as they are complex and lengthy. In this approach, each project has access to the appropriate form of management for its scope and uncertainty level.

Agile–Waterfall Combined Approach

One of the common tailoring approaches is to blend the Agile and Waterfall methods. Research by KPMG, conducted in collaboration with PMI, has shown that 60% of companies employed this combination of methodologies in 2022, reflecting an 11% year-over-year increase.[61] The Waterfall approach is a well-established framework that is the default for many companies, especially in industries prioritizing structure, predictability, and regulatory compliance. However, modern project demands often call for the flexibility and iterative progress of the Agile method. By blending these two methodologies, organizations can leverage the strengths of both approaches.

The Agile methodology primarily fits the needs of smaller teams and serves the software development industry. Multiple elements from this approach can still be applied to larger organizations using a hybrid model to build a flexible system. Taking advantage of the strengths of Agile, like faster feedback loops, helps teams to rapidly digest stakeholder feedback and adapt as needed over the project lifecycle. Identifying issues early reduces risks and increases customer experience satisfaction. In addition, the hybrid strategy significantly speeds up time to market and allows organizations to complete products and features on time.

Beyond these benefits, the hybrid approach adds visibility and responsibility. If team members operate in an open communication style, progress is transparently documented, and it is easy to identify the bottlenecks and inefficiencies. This adaptive prioritization means teams can deliver the most valuable features first and leave customer needs at the forefront.

However, the hybrid Agile paradigm has its downsides. One of the biggest downfalls of Agile is that, at times, it does not define the critical path or inter-project dependencies (which are often more evident in the Waterfall context). The hybrid approach solves this by incorporating formal components to help with planning and risk identification. Additionally, Agile frameworks have limitedly defined roles in the project lifecycle, which may not be sufficient to support complex, cross-functional projects in matrix organizations. Incorporating traditional frameworks can bring valuable stakeholder management practices to the hybrid model.

Full Agile implementation often includes creating a continuous delivery pipeline with many technical dependencies. Through the right balance between Agile and Waterfall control processes, the hybrid Waterfall–Agile method allows organizations to take advantage of the two methodologies to achieve better project results.

The following is an example of a model that combines Waterfall planning, including requirement collection and design, with iterative execution based on Agile Sprints. After the development is completed, the framework reverts to the subsequent deployment steps. This hybrid approach leverages the structured planning of Waterfall methodology and the flexibility of Agile for a more adaptive and efficient project execution.

Figure 8. Tailored Project Management Framework Combining Waterfall and Agile Methods

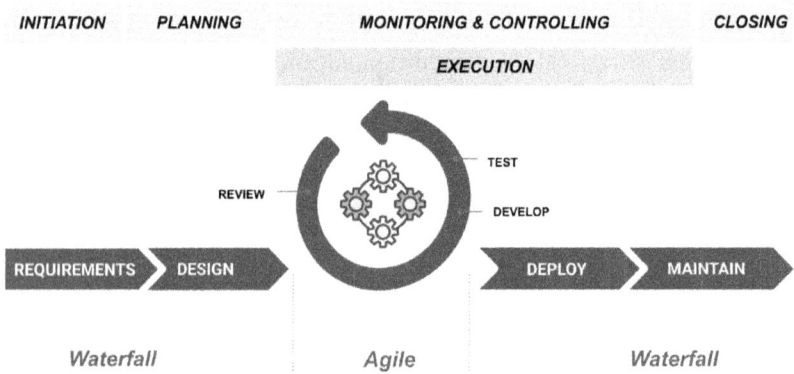

This type of tailoring allows project managers to optimize and prioritize tasks to deliver greater project effectiveness and relevance, ultimately helping drive better project performance by managing according to the project's needs.

Chapter 6

The Growth Mindset in Stakeholder Management

F ocusing on stakeholder management ensures the success and sustainability of a project or organization by connecting purpose and objectives, creating the trust and legitimacy that support ongoing partnerships. Making the process transparent for stakeholders guarantees that he whole team can collaborate productively and make better decisions, the latter because their input helps identify needs and priorities. By addressing risks early, organizations can minimize uncertainty and deal with obstacles ahead of time. In addition, it optimizes resources by focusing on the most important and impactful stakeholders and spurs innovation by combining different visions.

Strong stakeholder management is essential for meeting the requirements of the regulatory bodies and enabling responsible conduct that is necessary to bring success and positive impacts to the organization. In short, effective stakeholder management involves sharing values and goals through a supportive relationship.

Adaptive Stakeholder Analysis

Stakeholder analysis is the process of mapping and understanding the stakeholders (persons, organizations, groups, or institutions) who might be affected by a project. Its goal is to gather and assess data about stakeholders, as well as their needs, expectations, and influence levels, so as to resolve any potential issues early on in the project.

Stakeholder analysis employing an adaptive growth mindset can provide many advantages, bringing flexibility, cooperation, and, primarily, learning, allowing for a constant adjustment of the established interactions. It fosters trust and transparency such that everyone can express differences without fear of retribution, which creates stronger relationships and results in efficient problem solving. By assuming that issues and criticisms represent opportunities, teams are able to respond to changing requirements, take risks in the beginning, and evolve strategies as needed.

Adaptive stakeholder management creates a positive and constructive interaction across a project and consists of the following steps:

- **Identify stakeholders**: Determine who the stakeholders or organizations involved in the project are.
- **Prioritize stakeholders:** Stakeholders should be prioritized according to their importance or contribution to the project's success so that funds are focused where needed.
- **Engage with stakeholders:** Adopt targeted communication and coordination methods to engage stakeholders.
- **Manage stakeholders:** Coordinate competing needs, close gaps, and stay on track with project priorities.
- **Learn from involvement**: Collect feedback, report results, and learn from stakeholder interactions.
- **Optimize for the future:** Learn from past experiences to develop and improve future stakeholder management processes in a continuous loop.

The following is a visual representation of the adaptive stakeholder management process. This model enables project managers to deliver more agile and efficient solutions while building stronger and deeper relationships.

Figure 9. Adaptive Stakeholder Management Framework

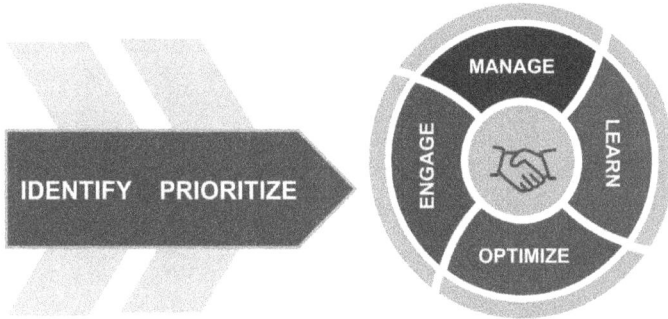

Utilizing the growth mindset-based stakeholder management model increases engagement and buy-in because stakeholders feel understood, involved, and included in the project outcomes, leading to more satisfaction and alignment with projects.

Stakeholders Analysis Techniques

Organizing project stakeholders is a fundamental step in understanding how different stakeholders influence and alter their involvement to suit the project manager's needs more effectively. The object is to discover what communications and procedures each group should use and which stakeholders must be accorded primary attention. Below are several techniques that may help analyze stakeholders.

Stakeholder Map: Power-Interest Grid

This method places stakeholders in a matrix of power (how much influence they wield over the project) and interest (how concerned they are for its success). The result is often a grid. In managing projects, project managers can use this grid to choose their priorities: those departments with both high power and high interest will require close management (personal approach, frequent interactions); those low on both counts may just stay informed (e.g., via a newsletter).

Figure 10. Stakeholder Map: Interest vs Power Grid

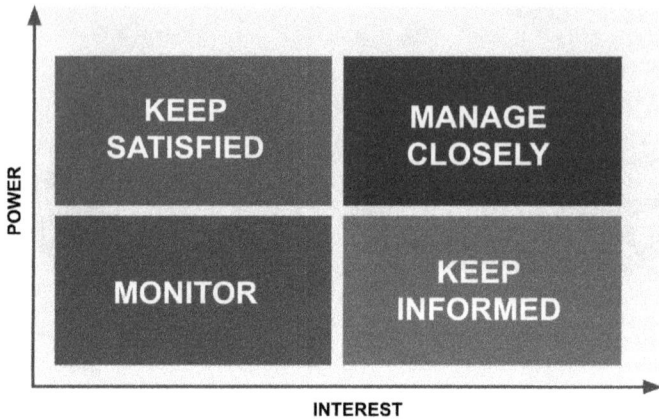

Stakeholder Analysis Matrix

The stakeholder analysis matrix provides a way to collect and process critical stakeholder information that can be used to support interactions and communications. It contains names, roles, interests and influence, needs, communication preferences, and engagement plans. Writing down these aspects allows project teams to identify stakeholder relationships better, predict potential problems, and seek partnership opportunities. It is a powerful tool for identifying custom-tailored initiatives to manage relationships and take proactive action on stakeholder issues.

Stakeholder Personas

Stakeholder personas are semi-realistic representations of specific stakeholder groups utilized to help teams know what they need, why they need it, and how they're facing a problem. Personas are developed based on interviews, benchmarks, surveys, and observation data and are typically given a fictional name, goals, discomfort points, power, affinity, and preferred model of communication. By humanizing stakeholder profiles, personas help project teams develop empathy and customize engagement strategies. They give more insight into stakeholder needs and ensure that messaging is valuable and practical.

The following table provides a comparison of each technique, incorporating growth-mindset factors that enable continuous learning and improved stakeholder engagement.

Table 12. Comparison of Major Stakeholder Management Techniques

Aspect	Stakeholder Map	Stakeholder Matrix	Stakeholder Personas
Purpose	Categorize stakeholders based on influence and interest.	Capture detailed, actionable insights on stakeholder roles.	Build user-focused profiles to guide engagement.
Format	Visual grid or map.	Tabular and structured.	Narrative profile format.
Best for	High-level prioritization.	Strategic planning and tracking.	Deep, personalized understanding and engagement.
Level of Detail	Medium.	High.	High.
Growth Mindset Aspect	Regularly reevaluate stakeholder categories as project dynamics shift; seek feedback to refine the map.	Use each interaction to validate or update assumptions about stakeholder needs and influence.	Continuous observation of stakeholder behaviors; defining personas based on real-world data.
Learning from Interactions	Monitor changes in influence or interest through consistent feedback loops; ask clarifying questions in high-priority areas.	Document insights from meetings or feedback sessions and refine action plans accordingly; clarify roles with stakeholders.	Use analytic tools, benchmarks, and structured interviews to gain deeper insights into stakeholder motivations and adapt the persona profiles.

Aligning Expectations with a Growth Mindset

Stakeholder expectations are the factors that determine whether a stakeholders consider a project successful. This assessment is based on both the deliverables and the overall collaboration between the project manager, project team, and stakeholders.

Controlling expectations begins with understanding what stakeholders want and why. Growth-mindset project managers welcome the challenge of the question: "But why are they asking?". They view expectations not as fixed demands but rather as dynamic needs that can evolve throughout the project. This perspective allows project managers to approach stakeholders with curiosity and openness, resulting in stronger collaboration. A growth mindset-oriented project manager solicits feedback, is problem-solving, and sees change as an opportunity to deliver more value. They use insights into the drivers of stakeholder needs to build empathy and relationships and devise plans to reconcile stakeholder needs with project requirements.

Below are some project management tools that can be of aid in working with stakeholder expectations:

- **The Project Charter**: This document provides project approval that details the business requirements, stakeholder perspectives, assumptions, and limitations. It also authorizes the project manager to begin planning.
- **RACI:** A chart that defines who is Responsible (R), Accountable (A), Consulted (C), and Informed (I), which helps define team members' responsibilities and their connections to deliverables, ensuring that expectations are clearly defined at the outset.
- **Project Scope Statement:** This document outlines a project's goals, scope, product acceptance criteria, and the configuration management needs to provide transparency to stakeholders.
- **Change Control Process:** A subsystem that defines the change process regarding deliverables and documentation, providing a systematic and repeatable approach to handling changes.

- **Project Communications Plan**: Document that defines requirements, communication channels, and timelines so the project manager can provide accurate and frequent updates throughout the project.

These tools enable transparency and communication and ensure that stakeholder expectations are understood, aligned, and delivered to achieve a project's success. To prevent misunderstandings and the establishment of unreal expectations, it is useful to identify roles and responsibilities early on so that everyone is aware of them before the project begins. Setting achievable objectives, establishing clear timelines, and outlining deliverables are essential to managing stakeholder expectations.[62] Finally, communicating progress regularly will help ensure that all the stakeholders' requirements are understood.

Managing stakeholder expectations with a growth mindset includes the following:

- **Knowing stakeholder needs**: Understanding what stakeholders want and why so as to be able to meet and fulfill their needs.
- **Building long-term relationships**: Investing in building trust and respect, not just in the context of the project but also for other future interactions.
- **Aligning on project outcomes**: Making sure stakeholders' needs are identified, documented, and included in the project objectives.
- **Identifying levels of engagement**: Determining how much of a role each stakeholder team will play during the project.
- **Encouraging stakeholder collaboration:** Getting stakeholders involved in decision-making and solving issues so they can provide useful feedback and feel a sense of ownership.
- **Clarifying communication preferences**: Using effective channels to deliver regular, meaningful updates.

- **Being open to changes**: Being open to shifts and having a defined system for suggesting, testing, and implementing changes within the project.
- **Managing conflicts:** Dealing with disputes early by debugging them and resolving them in alignment with the project's vision.
- **Welcoming feedback:** Receiving feedback regularly to incorporate it and make necessary strategic tweaks, as well as continually reviewing the strategy and stakeholder relationships.
- **Being proactive**: Drawing from feedback and experience to anticipate and build for the future, never waiting for issues to arise.

Selected aspects of the growth mindset stakeholder management strategy will be described in greater detail in the following sections of this book.

Applying the Growth Mindset to Stakeholder Communication

Stakeholder engagement is an integral part of any project's development. Given its complexity, stakeholder management is not limited to an individual project but is rather an evolving, long-term practice. Creating an effective communication strategy starts with identifying everyone who is interested in or has a bearing on the project and who will be affected by its results. It is also common in a project management practice to create a map reflecting the stakeholders' classifications and use it to tailor messages to suit their needs and desires.

After setting objectives, the next step is to outline the communication plan, starting with choosing the most relevant communication methods. The PMBOK® Guide states that communication with stakeholders can be verbal or written and received through formal or informal channels. The method employed depends on two interrelated factors: what kind of information is being released and how important this is to the target audience.[63] Besides that, PMI categorizes communication methods into three types: push, pull, and interactive communication.

Push communication is a one-way form of communication through which a company or organization actively seeks stakeholder involvement by sending emails, announcements, status reports, etc. Conversely, pull communication implies that a project's participants search for information. For example, we might find team members logging into an intranet site to look up corporate communication policies and templates; this helps to gauge stakeholders' concerns indirectly. Finally, interactive communication includes exchanging information with one or more stakeholders. Examples of this include meetings, brainstorming sessions, phone calls, chat conversations, and presentations.

Involving stakeholders includes relying on soft skills such as active listening, interpersonal skills, and conflict management, as well as leadership skills like the ability to craft ideas and think critically. Another layer would be to incorporate a growth mindset into this process. The table below presents some examples of applying it in practice.

Table 13. Tools for Applying the Growth Mindset to Stakeholder Communication

Tools for Applying the Growth Mindset to Stakeholder Communication		
Tool	**Type of Communication**	**How a Project Manager Can Use It**
Email	**Push / Written**	Open the dialogue and invite team members to ask questions, brainstorm, and share ideas.
Presentation	**Push / Written**	Make the presentation a way for the team to collaborate, learn from each other, and assess progress. Use language that focuses on growth.

Tools for Applying the Growth Mindset to Stakeholder Communication		
Progress Report	**Push / Written**	Stay development-focused, celebrate achievements and failures, and foster constant learning and improvement.
Centralized Repository	**Pull / Written**	Create a SharePoint site where stakeholders can access project documents, templates, and lessons learned.
Regular Meetings with Stakeholders	**Push / Verbal**	Schedule review meetings for stakeholders to present them work, discuss challenges, and share strategies for overcoming obstacles. Ask for feedback.
Knowledge Transfer Sessions	**Push / Verbal or Written**	In case a key stakeholder takes an extended leave, organize a session to share critical insights, methodologies, and undocumented processes.
Communities of Practice	**Push / Verbal or Written**	Establish a community for key stakeholders to proactively find solutions before problems occur and define areas to improve processes or refine strategies.
Collaboration Tools	**Push or Pull / Written**	Utilize tools like Slack, Microsoft Teams, or Webex to create dedicated channels for different project aspects, facilitating updates and feedback in a structured manner.
Post-Implementation Reviews	**Push / Verbal**	After the product launch, gather stakeholders and project team to discuss successes, challenges, and areas for improvement.
Post-Mortem Storytelling	**Push or Pull / Written**	During project retrospectives, encourage team members to share stories about effective problem-solving, embedding valuable insights into their practices. Document and make available for stakeholders to view.

Tools for Applying the Growth Mindset to Stakeholder Communication		
Establish Awards	**Push / Verbal**	Introduce "Growth Mindset Champion of the Month" award to recognize individuals who put effort into making meaningful improvements. Make stakeholders a jury to choose the winner.
Cross-Training	**Push / Verbal or Written**	Use training tools to build awareness among stakeholders about the key deliverables and metrics being used to measure success.

Continuous Feedback Loops

Feedback loops in this context refer to the process of gathering, processing, and acting on information (usually with regards to performance or results) in order to optimize or modify it continuously. When it comes to project management, feedback loops involve continually checking the progress of a project with stakeholders, team members, or customers. This information is then used to plan further adjustments, process optimizations, and improvements in delivery quality. Through continual review of the project at various stages, feedback loops enable iterative problem-solving and decision-making, keeping the project on target and flexible to change.

Feedback loops are closely connected to a growth mindset, or the view that capabilities and results can be learned and improved over time. A growth mindset makes it a goal for employees and teams to perceive feedback not as criticism but as a learning opportunity. Incorporating feedback as a part of project management fosters continuous improvement, creating a culture in which problems are treated as opportunities for learning. Feedback loops fuel the growth mindset by confirming that each bit of information received feeds forward to team members' personal and professional success.

The project manager contributes to creating feedback loops by gathering, analyzing, and using feedback throughout the project. They are responsible for creating an open communication platform, allowing stakeholders to share their views at the right time.

There are several methods project managers can use for soliciting stakeholder feedback that are suited for different phases and contexts of the project:

- **Individual meetings** with stakeholders are helpful when a project manager needs to receive specific feedback and ideas.
- **Weekly check-ins and progress meetings** provide a valuable tool for iterative real-time feedback so that stakeholders can share their concerns or feedback as the project progresses.
- **Surveys, polls, and questionnaires** are quantitative feedback tools that provide a generalized sense of stakeholder views.
- **Workshops or brainstorming sessions** allow stakeholders to solve problems creatively and generate suggestions, typically resulting in valuable insights.
- **360-degree feedback** is a comprehensive method whereby feedback is collected from all directions (team members, managers, stakeholders, clients) to provide a holistic view of how the project is performing and where improvements are needed.
- **Informal touchpoints** with stakeholders foster personal relationship-building and facilitate the gathering of more data.
- **Indirect methods**, such as observation or tracking instruments (like performance metrics or UX data), can also gather feedback, showing how the project is received beyond the realm of direct interaction.

These methods ensure that stakeholder insights are continuously integrated into a project's design.

Types of Feedback

When giving feedback, it is crucial to ensure that it comes from the right channel and in the correct format. While a variety of feedback types exist, not all of them are appropriate for all situations. For instance, immediately after a significant success, such as conducting a productive training, a team member may prefer praise-based feedback about how things went well rather than constructive feedback on how things could be done better. The constructive kind of feedback is helpful, but it may not be suitable immediately following an event when emotions are still fresh. Below is a brief description of the seven types of work-based feedback:[64]

- **Appreciation feedback** lets team members know they are appreciated and gives them a boost. Specific praise, such as acknowledgment that a difficult deadline was reached or a difficult project was completed, is better than standard praise because it is more personal and meaningful.

- **Formal feedback** is received in an organized environment—for example, in a performance review—and it provides a very specific picture of what and how employees are doing. It reviews progress toward objectives and allows team members to know what is being worked on and what still needs work.

- **Guidance feedback** combines praise and advice so leaders can make kind suggestions in a non-aggressive way.

- **Motivation feedback** improves morale when the team is in trouble or when the project encounters risks. It is instantaneous, and it provides comfort with a few words of encouragement to maintain workers' confidence.

- **Forward feedback** targets future growth rather than the previous error, thereby motivating individuals to reach their objectives. It focuses on the "why" and lets people see their part in company goals.

- **Coaching feedback** is less formal and more strategic in nature. Senior project managers are the coaches, combining advice and follow-up in the reviews to motivate the team to improve.

- **Informal feedback** is casual; comments are involuntary and come with the territory of everyday life. This does not involve having meetings but rather refers to the providing of quick praise or input at everyday times, such as a brief talk in the hallway or after a meeting.

Understanding, using, and applying feedback models will allow the project manager to accommodate all team members' needs and ensure an effective, inclusive, and positive project culture. Providing feedback effectively is also an essential aspect of managing stakeholder communication, analyzed in the previous section.

Managing Conflict with a Growth Mindset

Conflict is an inevitable aspect of all relationships, both personal and professional. It can provoke stress, frustration, and anger but also represents an opportunity for growth, collaboration, and innovation. Workplace conflicts are real and can significantly influence project employees' well-being, engagement, and performance.

Statistics published by the Society for Human Resource Management (SHRM) indicate that incivility is not uncommon in the workplace—almost two-thirds of U.S. employees can claim that they have experienced a variation of incivility on the job.[65] The result of this is a massive reduction in job satisfaction, as workers who describe their work environment as "uncivil" are three times more likely to be dissatisfied with their job than those who do not and are twice as likely to leave within the next year. Given the fact that these tensions have continued to grow, it is alarming that one-third of U.S. workers are prepared for workplace conflict to worsen in the future.

Compounding these problems, according to Gallup, U.S. employee engagement is declining, with just a third of employees saying they're engaged at work.[66] This reflects a slight drop in actively disengaged employees from 18% in 2022 to 16% in 2023. This translates into the waste of approximately 18% of yearly salaries due to unengaged employees and the organizational cost of disengaged employees, which reaches between $450 and $550 billion annually.[67] The most significant of these are escalation costs, which are associated with terminating the employment relationship, particularly when formal processes or illnesses are involved.[68]

Workplace conflicts, often rooted in personality clashes, egos, and stress, further exacerbate these challenges. Nearly half of all workplace conflict stems from personality differences (49%), with significant contributions from stress (34%) and heavy workloads (33%).[69] Yet, in the face of these challenges, nearly half (55.7%) of employees put restoring harmony above all else.[70] This shows how much room there is for the management of conflicts using growth mindset tools, which is highly relevant in the context of the project management landscape that we are outlining in this book.

A project manager's perception and response to conflict depends on their mindset. With a positive and growth-oriented mindset, a project manager views conflict as a platform for learning, development, and empathy rather than as a threat, setback, or personal affront. Additionally, in the context of growth mindset project management, conflict may even create room for practicing effective communication, active listening, and the pursuit of constructive resolutions, as opposed to avoidance, blaming, or escalation.

The effective and constructive navigation of conflicts is certainly one of the abilities of a growth mindset project manager. The first step in practicing it is approaching the situation by reframing the issue positively, viewing it as an opportunity, not as a challenge. Rather than shying away from conflict, courageously and confidently engage with it.

Another growth mindset tool that helps with conflict-solving is listening attentively to the other party's standpoint, identifying common ground, and striving for mutually beneficial solutions. Following conflict resolution, reflecting on the lessons learned and soliciting feedback from the other individual is one of the best practices to engage in as well.

Conflicting situations test one's ability to maintain a growth mindset approach. In a mindful way of managing conflicts, it is crucial to monitor self-talk and remain attentive to negativity. From a broader perspective, creating a growth mindset environment helps to reduce the room for conflicts. A growth mindset project manager can create this environment through leading by example, surrounding themselves with individuals who share the same mindset, and boosting each other's morale to reach objectives. Ideally, this can be applied to a project management team, but there are also individuals outside of the immediate team who can help expand a growth mindset network. Engaging with communities or groups that endorse learning, growth, and cooperation for added reinforcement helps to maintain a positive way of thinking and to manage conflicts efficiently.

Managing a conflict situation with a growth mindset is not an easy task, and it may require leaving one's comfort zone, which is why it is important to approach the situation in a mindful way from the beginning to the end.

Handling Difficult Conversations

The workplace resource startup Bravely found that 70% of employees avoid difficult conversations.[71] The fear of retaliation and negative relationship impacts, as well as a simple lack of training, causes staggering numbers to avoid engaging in challenging dialogues. Statistics reveal that the average cost of any failure in conversation to an organization is $7500 and that over seven working days are wasted. 53% of employees deal with "toxic" situations by pretending they don't exist.[72] This leads to the erosion of employee engagement and trust in the organization. The data shows that being anxious about engaging in difficult conversations is not uncommon. For many people, difficult conversations are associated with danger and ambiguity. Embracing the unfamiliar in our conversations and lives may conflict with our sense of self, our principles, and our aspirations. Difficult conversations are a typical project management pain point; their cause is multifactorial.

Conflicting priorities or expectations between team members or stakeholders leads to misalignment, and unclear communication usually brings uncertainty. Tension around performance issues, including underperformance or lack of accountability, can also arise, particularly when protecting deadlines or quality. Collaboration renders the situation even more complicated, as does scope creep—changes to the direction of a project cause confusion. Limited resources, time constraints, and budget pressures always lead to difficult conversations about what must take priority. Moreover, the inadequacy of schedule estimates on large projects can force tough reassessments when facing unanticipated risks or project setbacks.

On top of that, any leadership challenge or emotional reaction to the situation—including stress, the fear of failure, the fear of one's own position and perception, individual perspective based on one's own experiences, the "what ifs" associated with the situation—and the specifics of an issue can hamper conversations. Legal or company internal compliance policy limitations can complicate project conversations even more, particularly those involving third parties (vendors and clients). Nevertheless, these conversations are critical to ensuring that projects continue to make progress and team spirits remain high. Several growth mindset skills can help handle these challenges and maintain the project on the right path.

Framing Challenging Conversations with a Growth Mindset

If a conversation is labeled as difficult, it is likely to trigger an individual's nervous system into action. The harboring of a growth mindset can break this cycle by framing it more positively instead. Experts say that a tough conversation is best handled by simply treating it as an everyday conversation.[73] The way we perceive the conversation is reflected in the language we use. The table below includes examples of some challenging project management conversation statements with their growth mindset alternatives.

Table 14. Shifting from Fixed to Growth Mindset Language in Challenging Project Management Conversations

Fixed Mindset Language	Growth Mindset Language
The deliverable was not delivered.	*We failed to meet the standards for this deliverable. Let us review how we can do better in the future.*
There is not much time for this.	*You know, we have a tight deadline, but we could be efficient if we just focus on priorities.*
The communication in the team is problematic.	*The team has an opportunity for improvement around communication—something that we can solve with the right tools or processes.*
We have limited resources.	*We have some resource constraints, so we'll need to prioritize key deliverables.*
We missed the deadline.	*We missed the deadline, but there is a learning experience here that we can build on to improve planning for the next phase.*
The project requirements are not sufficient.	*Stakeholders want the project requirements to be clearer.*

Even if things do not go according to plan, there is always a way to consider what can be learned from the situation. This approach is beneficial from the perspective of creating partnerships among the team members and stakeholders as it ultimately fosters a growth mindset culture. This involves creating a continuous improvement approach to encourage everyone involved to continue learning and adjusting. Personal conflicts are re-framed as developmental opportunities so that discord is transformed into better communication and stronger teamwork. This mindset encourages growth, builds stronger relationships, and enhances the team.

Encouraging collaboration and open dialogue will help address project challenges. When team members are not afraid to speak their minds, they can brainstorm more ideas and work together more effectively to find viable solutions. Fostering this atmosphere of open dialogue allows for different inputs to be considered while also offering conflicting parties the opportunity to come together and find a solution as a unit. Assuring transparency creates trust, thereby allowing everything to be monitored clearly and work to be carried out in a collaborative fashion, ensuring accountability and engagement.

Promoting constructive feedback and a continuous improvement mentality will lead to the development of leadership style as teams blossom over time. Project managers can ensure that the input they receive is transformed into actionable insights and recommendations for their team, thereby helping to improve processes or performance and nurturing growth. Additionally, when feedback becomes a learning opportunity and not criticism, it leads to improvement and empowers the team. Maintaining this habit of improvement gets the team excited and motivated to obtain better results.

Fostering an environment in which team members see their failings as a chance for development, are willing to accept feedback and change, and focus on collaboration rather than blame only works if project managers implement growth mindset tools during challenging conversations. Growth mindset project managers also develop rituals for celebrating growth, which can be adopted by celebrating progress and applying what you learned to future situations.

Chapter 7

The Role of AI in Transforming Project Management

The global AI revolution is undeniable—the acceleration of Artificial Intelligence adoption is transforming industries worldwide. In October 2024, 75% of companies were experimenting with or deploying AI, according to the *2024 Work Reimagined* research conducted by EY, which surveyed almost 1600 employees across 23 countries and 27 industries. It is worth mentioning that this number increased dramatically from the 22% recorded in 2023.[74] About 70% of CEOs surveyed by PwC[75] in 2024 say that generative AI will fundamentally transform how their organizations produce, deliver, and extract value. They seem to believe in both the speed of generative AI adoption and the speed of its implementation.

AI: The Next Global Revolution

The rise of Artificial Intelligence represents a transformative shift on a scale that may one day be compared to prior historical revolutions, such as the Industrial Revolution or the Digital Revolution, which totally reshaped economies and global labor markets. As AI technologies advance, they are set to revolutionize almost every sector—if not all—ranging from manufacturing, construction, and transport to retail, finance, healthcare, energy, legal, and education. At the heart of these changes is the technology industry.

In the article "GPTs: An Early Look at the Labor Market Impact Potential of Large Language Models"[76] by Tyna Eloundou, Pamela Mishkin, Sam Manning, and Daniel Rock, the authors examine the potential implications of Large Language Models (LLMs), such as GPTs, on the U.S. labor market. The study finds that LLMs could impact approximately 80% of the workforce. Around 19% of workers may see at least half of their tasks impacted. Higher-income jobs are expected to experience greater exposure to LLMs and LLM-powered software, and the effects are not limited to industries with higher productivity growth. The research also indicates that with LLMs, about 15% of worker tasks can be completed faster without compromising quality, and this share increases to 47-56% when considering LLM-powered software. These findings suggest that LLMs, as general-purpose technologies, may lead to significant global economic, social, and policy changes.

PwC[77] estimated that AI's expected contribution to the global economy by 2030 may be worth $15.7 trillion. This will be achieved by product enhancements, which will stimulate consumer demand by increasing product variety, personalization, and affordability. As noted in PwC's *Global Artificial Intelligence Study: Exploiting the AI Revolution* report, the greatest economic gains from AI will be in China (a 26% boost to GDP by 2030) and North America (a 14.5% boost), equivalent to a total of $10.7 trillion and accounting for almost 70% of the global economic impact.

Artificial Intelligence is currently not only the subject of economic analysis but also the subject of discussion by global political leaders, which only reinforces the notion that we are witnessing a true worldwide revolution. At the World Economic Forum's Annual Meeting in Davos, 2024, "Artificial Intelligence as a Driving Force for the Economy and Society" was a key theme.[78] AI is now recognized as a powerful tool for addressing global challenges, and there is no doubt that it will change the economy as we know it. Leaders, however, have stated that its potential requires both innovation and regulatory frameworks.

Transformation of Project Processes and Outcomes

The AI revolution is profoundly reshaping the project management field, influencing how projects are planned, executed, and managed and thus altering the roles and responsibilities of project managers. The process started with early AI, with machines performing simple tasks, but it has since undergone a chain reaction, leading to a boom when tools like ChatGPT emerged in 2022 to pioneer generative AI.

AI-enabled tools are now being more and more often integrated into project management platforms to help with tasks such as data analysis, risk mitigation, and resource management, which significantly boosts efficiency and decision-making. For project managers, AI can be extremely useful. It can automate repetitive tasks and make complicated decisions backed by powerful data analysis. AI enables project managers to foresee risks and improve timelines and communication, transforming their roles into more strategic positions. This, in turn, implies a transition towards data-driven, more informed project execution.

The Project Management Institute's 2023 research revealed that 91% of the surveyed project management professionals believed AI would have at least a moderate impact on their profession, with 58% describing the impact as "major" or "transformative."[79] Additionally, 82% of senior leaders in the project management field anticipate AI will significantly influence how projects are managed within their organizations over the next five years. This data underscores the growing role of AI in reshaping project management and general business management practices.

PMI also outlined a three-tiered model to evaluate how AI transforms project management roles and practices. This framework highlights distinct areas of AI's influence, tailored to the complexity of tasks and the degree of human intervention required. The tiers are as follows:

- **Automation**: Generative AI automates manual tasks such as producing reports, analyzing meeting minutes, and performing calculations. This enables project managers to streamline processes and outsource everyday tasks without going through an extensive review.
- **Assistance**: Generative AI supports experts at this stage by preparing preliminary analyses, schedules, and risk evaluations. Such outputs must be moderately polished and managed by senior managers to be accurate and meaningful.
- **Augmentation**: Generative AI adds capabilities for complex strategic tasks like building business cases and navigating complex decision-making processes. AI becomes a co-worker here, while the project managers take advantage of their knowledge to iterate and steer the effort.

KPMG[80] provides more specific insights on AI in the world of project management, signaling key areas where the project manager will benefit from applying AI:

- **Informed decision-making**: AI's data analytics capabilities identify trends and critical information in real time, helping project managers spot risks and recommend corrective actions instantly.
- **Management of resources**: AI helps track resources, ensuring projects stay within budget and on schedule by recommending optimal resource allocation and identifying schedule conflicts.
- **Automation of repetitive tasks**: AI automates routine tasks such as reporting and scheduling, enabling project managers to focus on higher-level strategic planning and leadership.
- **Real-time access to information**: Advanced AI tools provide project managers with instant access to project data on the go, enhancing productivity and enabling better decision-making, no matter their location.

It is still relatively early to be able to predict the full impact that AI transformation will have on the project management field; however, it will undoubtedly be significant. The results we have seen in the last couple of years of incorporating AI into project management are encouraging. Not only does AI help manage key processes and improve the quality of deliverables, but it also helps ensure that projects are better aligned with organizational goals and strategies. AI-driven workflows enable organizations to complete projects faster and more efficiently thanks to productivity savings and smaller budgets.

The ways project managers work are also changing. Artificial Intelligence is being used to automate and add value to various steps of the process, from planning to time and cost estimation to risk assessment. This gives teams more time to innovate and solve problems. With the increasing usage of AI, project managers will continue to discover how to perform better on projects, enabling long-term growth for businesses.

AI will also certainly change the way project managers learn and develop. Gathering skills may be easier and more "democratic"; training and interactive learning are available virtually for everyone. On the other hand, AI may require many project managers to adjust to the new reality and be more tech-savvy.

Leveraging AI to Cultivate a Growth Mindset

The rapid progress of Artificial Intelligence aligns with the principles of a growth mindset. By fostering continuous learning, innovation, resilience, and collaboration, AI helps individuals and teams positively welcome change. AI solutions based on machine learning can offer actionable insights that reflect the iterative essence of a growth mindset, allowing project managers and teams to correct mistakes and tweak their strategies.

The AI Revolution in Learning and Development

The incorporation of Artificial Intelligence is revolutionizing the traditional ways of learning and development by introducing innovative solutions for employees. This technological transformation impacts skills enhancement, knowledge retention, and overall performance outcomes[81]. Below are some examples of ways to leverage AI to transform employees' traditional learning experience:

- **Tailored learning paths:** AI-driven machines scan learners' habits and preferences to deliver personalized training. This personalization ensures employees receive timely and impactful learning to get the best ROI for their development initiatives. Creating a personalized learning plan is a top priority for 71% of HR executives to rapidly close the skills gap, per research conducted by the IBM Institute for Business Value.[82]

- **Intelligent learning platforms**: AI helps the platform modify the content as learners learn and move forward. They give immediate feedback, coaching employees on where they are lacking and where they are doing well. This live and individualized method enhances participation in training by serving personalized learning demands.

- **Enhanced performance evaluation**: AI data allow for real-time employee tracking to detect skills shortages and guide individual interventions. Through predictive functionality, companies can already align training with the requirements of a future workforce. This scientific mindset builds an attitude of continual learning and ensures that L&D initiatives are focused and efficient.

- **Learning collaboration and knowledge sharing:** AI can enable learning environments in which employees are connected to other learners with the same learning objective or common skills. Employers can learn from each other, get mentored, and solve issues collectively with the help of AI-powered platforms—all to encourage a culture of learning and innovation within the company.

- **Gamification and engagement**: AI can add gamification to corporate training to increase engagement and motivation. With customizable challenges, rewards, and tracker tracking, AI makes learning fun and engages participants. Gamification also lends itself to competition and cooperation, which drive participation.

- **On-demand chatbots**: AI chatbots offer 24/7 learning support and training to make the process of training simpler. They give personalized suggestions, answer questions, and send employees the right content to learn and develop efficiently.

Enhanced Collaboration

AI-enabled collaboration tools are also changing how teams work by creating platforms for joint contributions, decreasing inefficiencies, and making it easier to interact with each other. AI-enabled solutions such as Microsoft Teams, Slack, Zoom, Jira, Asana, Miro, Figma, Canva, and many more offer active capabilities that facilitate communication, task management, and workflows, reshaping the end-to-end collaborative process. To give a few examples of this change, the Zoom AI Companion, besides recording meetings, allows for summarizing conversations and assigning action items. Slack's AI-based priority tool highlights and prioritizes high-value conversations, preventing teams from falling behind on information and enabling them to get back to business. Asana's AI Work Graph® powers customized workflows, triages risks, manages action projects, and automatically reallocates work.

These advances dramatically change the nature of teamwork by promoting transparency, accountability, and speed to achieve project goals faster. Insights and predictions in real-time encourage proactive cooperation, such that team members can identify needs, react, and modify strategies together. By making teams more efficient and cohesive, AI-powered platforms are not merely tools—they are also participants in a team's performance.

The Nielsen Norman Group[83] conducted interesting research on the use of generative AI tools in business. The results, averaged across three studies, revealed that they improved users' performance by 66%. They also showed that more complex tasks included more significant gains and that less-skilled workers benefit the most from AI use. The specific results of the study were as follows:

- *Study 1*: Support agents who used AI could handle 13.8% more customer inquiries per hour.
- *Study 2*: Business professionals who used AI could write 59% more business documents per hour.
- *Study 3*: Programmers who used AI could code 126% more projects per week.

While AI collaboration tools have numerous benefits, they also come with their own challenges. Employees are often excited about the possibilities that productivity tools provide, forgetting about the potential threats. The number one concern is data security threats, as these platforms hold information about projects that, if compromised, could result in major losses to businesses. Additionally, excessive AI usage could also reduce a team's critical thinking and problem-solving abilities as employees become overly accustomed to the suggestions of automated tools. Other risks arise from the biases built into AI algorithms, which can unintentionally engender disparities in assignments or decision-making. More insights regarding concerns about AI adaptation can be found at the end of this chapter.

Fostering Innovation

AI empowers project teams and product designers to create new and creative solutions by looking at large amounts of data and identifying market trends. AI can quickly search for patterns in past data and analyze historical information and current tendencies that can guide future processes and product enhancements. Teams can use Artificial Intelligence to forecast changes or find market opportunities that can inspire creative ideas.

AI simplifies research and product comparison by unifying insights from multiple sources. It allows project teams to capture and visualize market information and competitors' activities and to understand customer expectations. Learning from others makes it easier to stay ahead by adapting quickly.

Through advanced AI-assisted benchmarking, project managers can avoid wasting time trying to reinvent the wheel. AI models, including recommendation engines and natural language processing algorithms, help teams choose the best ideas from suggestions received. These tools can parse massive amounts of internal and external data to find which ideas or functionality will most likely appeal to customers and market demand. Teams can steer away from making costly mistakes and concentrate on initiatives that have the best chance of success if they prioritize ideas based on data.

AI also helps teams to innovate by building and simulating product scenarios in a virtual environment before offering them to real customers. AI-driven simulation tools can model the behavior of a new product or service under various conditions and give insight into the risk factors and potential customer reactions. This "sandbox" approach enables teams to test their products without the financial or reputational risk of real-world trials. By enabling experimentation, AI helps to develop a growth mindset culture in project teams, where the team feels comfortable trying out and finding innovative solutions.

Overall, AI fosters product and project innovation by spurring creativity, enabling research, developing concepts, running experiments, and creating an environment of innovation and teamwork.

Shifting from Fear to Opportunity

Shifting from a perspective of fear to one of opportunity is vital to successful project management as it promotes adaptability and continuous team development based on growth mindset principles. AI, as a rapidly emerging trend, may be received as a threat by many project teams. On the project level, however, it can open up possibilities for collaboration, efficiency, and risk reduction. Seeing this as an opportunity will allow project managers and their teams to approach technological changes with a positive, forward-thinking perspective. In that way, potential challenges are transformed into avenues for growth.

At the personal level, individual members of a team can also significantly profit from AI's ability to access information streams and make tasks more inclusive. For example, many expat workers employed in foreign markets struggle with their writing skills, which is a common issue for foreigners from all over the world working in the tech industry in the United States. AI writing tools, e.g., Grammarly (AI-powered writing assistant), support such professionals by offering them advanced language aid. These tools can recommend grammatical corrections and format improvements and even provide live translations, enabling individuals to convey their message with greater facility. By simplifying language and facilitating communication, AI creates a more inclusive workplace.

AI-Driven Decision Support

Decision-making support is inherently tied to the concept of a growth mindset, which stresses enhanced learning, increased resilience, and improved problem-solving. In project management, this mindset is crucial, as decision-making is often based on evolving information, changing conditions, and the need to adapt strategies to achieve successful outcomes. AI supports this mindset by giving project managers the tools needed to make decisions, predict future outcomes, and optimize project performance.

AI-driven decision support systems are designed to aid project managers in taking informed action by analyzing vast amounts of data, identifying patterns, and providing on-demand, actionable insights. With the help of AI, a decision can be made not based on intuition or partial data but on complete, up-to-date information. The advanced forecasting capabilities of AI play a significant role in modern decision-making processes by allowing project managers to quickly anticipate future outcomes using past data. A recent Deloitte report shows that organizations leveraging this type of analytics can boost project success rates by almost 75%.[84] Early monitoring allows teams to detect risks, such as cost overruns or lags, so that they can mitigate them and spot new trends or opportunities to exploit. In addition, AI can generate accurate timeline forecasts to provide better project planning. Key data-related areas that AI supports are the following:

- **Analysis and data insights:** AI can look into previous and ongoing project data and identify patterns and irregularities to help project managers make evidence-based decisions.
- **Real-time analytics**: AI systems can crunch information in real-time so that managers can change course immediately based on live performance data.
- **Objectivity and accuracy:** AI removes human judgment from decisions, thus making them solely based on facts, which creates objective project management.

The other domain in which AI can make decisions is resource allocation and optimization. AI maximizes utilization by monitoring the skills, workload, and project requirements of members of the team and rendering real-time task-splitting recommendations. It coordinates work with the best talent and improves productivity and resource effectiveness. AI can detect possible resource scarcities so that managers can rework or request more before the over-abundance of resources develops.

Artificial Intelligence can be implemented early in the resource management process and applied across the entire end-to-end HR process, extending beyond the project management domain. Predictive analytics, powered by machine learning, enhances HR functions such as hiring, retention, and succession planning by consolidating and analyzing scattered data.[85]

AI can also help with decision-making through simulation scenarios, which enable project managers to analyze possible outcomes and determine various strategies. After adding variables, AI computes several possible situations, allowing project managers and their teams to see how different options can impact the project outcome. These simulations provide helpful knowledge about risk reduction, enabling the measurement of the impacts of different risks and the making of plans to counter them before they happen. Lastly, AI assists with the strategic decision-making process by suggesting the most optimized solutions.

Finally, AI enables the proactive monitoring and control of projects, allowing for immediate visibility into project outcomes. By analyzing real-time data, algorithms provide real-time feedback to project managers on key performance metrics like budget compliance, work completion, and output. AI also automatically triggers notifications when project metrics fall behind the plan to enable taking corrective actions. During project execution, it constantly updates its suggestions as KPIs become available, ensuring that a team's strategies remain adaptive to the current situation and optimizing the overall project success.

The following table summarizes how AI can support the decision-making process and connects it to the implications of the growth mindset.

Table 15. Selected AI Applications in Project Management

Area	AI Application	Impact on Project Management	Growth Mindset Implications
Decision Support Systems	AI processes data sets to provide insights and trend-based recommendations for actions.	Enhances strategic decision-making by offering faster, more accurate recommendations.	Enables learning from data-driven insights, developing continuous improvement.
Resource Allocation & Optimization	Machine learning algorithms optimize resource distribution by predicting shortages and overages, improving efficiency.	Leads to cost savings and better utilization of resources, preventing project delays and budget overruns.	Emphasizes continuous improvement and adaptability.
Scenario Simulations	AI models "what-if" scenarios to predict potential outcomes under various conditions, helping teams prepare for uncertainties.	Improves risk management and allows for proactive adjustments to project plans.	Encourages adaptability by exploration of different approaches and understanding that failure is an opportunity to learn and adapt.
Project Monitoring & Control	Real-time AI-powered dashboards track project progress, flagging issues and recommending corrective actions.	Increases project success rates by enabling timely interventions and adjustments, leading to higher on-time delivery rates.	Fosters resilience and adaptability by framing challenges as opportunities to course-correct and improve processes continuously.

Advancing Careers with AI in Project Management

Artificial Intelligence is now an integral part of project management and is changing the way project managers work and develop their careers. As AI transforms project management processes, project managers must keep up by learning and leveraging new technologies and growing as AI transformation leaders. Often, they are the ones best positioned to implement AI in their organizations. The AI revolution allows project managers to grow but requires them to extend their knowledge and, often, gain new skills.

The AI transformation is increasing the need for project managers to be trained and to stay ahead of the curve. It is expected that project managers will learn the different new tools being offered on the market at a super-fast pace. This includes tools for managing projects, such as AI-driven applications for generating and optimizing project plans, automating repetitive tasks like report generation and scheduling, managing risks through predictive analytics, visualizing real-time data, automating resource allocation, tools for creative writing, image development, and presentation creation, as well as improving communication, among many others. On top of that, project managers are encouraged to understand AI algorithms, prediction models, and data analysis tools.

The future of project management is being recast not only in technical terms but also as a project manager able to think in new and different ways. By automating everyday activities, AI frees up time for decision-making at a high level and problem-solving creativity on a larger scale. Project managers can model scenarios, analyze risks, and experiment with new solutions using AI-based simulation software. This mental change is aligned with the growth mindset, which is a mindset in which all changes can be seen as opportunities. When project managers adopt AI, they are advancing their careers and those of their companies.

While AI is transforming the technical aspects of project management, the human element remains essential; the important thing is not that AI tools produce useful information but rather that the project manager makes sense of this information and puts it into action. Emotional intelligence, conflict management, and team motivation are all skills that a project manager must nurture to lead teams and make the right decisions, as they are responsible for guiding them through the AI landscape.

The future will be dominated by AI, and careers for project managers will be thriving if they adapt to this reality. Since AI is going to become more and more prevalent in the context of project management methodologies, if you are ahead of the game and stay up to speed with all things AI, you will be in a good place for your career.

Project managers can do so much more with these AI technologies to ensure they lead projects that run smoothly. As tools continue to improve, project managers who combine technical knowledge with leadership and planning will be the most in demand. The AI-first shift in project management is not just about automation but is rather about how AI will complement humans and create a greater impact.

Challenges of AI Transformation

Bryan Robinson, the author of *Chained to the Desk in a Hybrid World: A Guide to Balance,* states in his article in Forbes on the challenges of AI implementation that "The rise in AI has caused a mixture of excitement and fear as it becomes a standard part of our lives. Some say it's a risk to civilization as we know it, while others insist it will transform the way we work, live, and interact with one another. The part that is not debatable is that not only is AI not going away, it's on the upswing. You can get that tattooed. So, it's important to develop a reasonable comfort around its use."[86]

Although the AI revolution is inevitable and brings many positive implications, there are also some threats and challenges that need to be addressed. In response to these concerns, new solutions and regulatory standards are needed at the industry, state, and global levels.

Employers Expectations

Using AI can sometimes result in employee reluctance, as some members of the team feel disenfranchised due to a lack of awareness of the technology or ability to change their way of working. In addition, employers have high expectations tied to AI-related promises of productivity savings. In his Forbes article, Bryan Robinson analyzes the findings from a study by The Upwork Research Institute that interviewed 2,500 global C-suite executives, full-time employees, and freelancers. The results showed that despite 96% of executives expecting AI to improve productivity, 77% of employees who use AI find that it increases their workload and presents challenges in achieving the anticipated benefits. A significant number of employees feel overwhelmed, with 47% unsure how to meet their employers' productivity standards and 40% feeling that their employers are asking for too much. This additional pressure has resulted in burnout, with 71% of full-time employees experiencing it and 65% struggling to cope with the heightened productivity demands the situation creates.

This research highlights the importance of using AI tools thoughtfully and with appropriate training, setting realistic expectations, and maintaining a strong sensitivity to workers' ability to manage change. Adopting AI in the workplace requires addressing employee concerns and preparing them for these changes to avoid burnout and frustration.

Data Concerns

Data security issues related to AI tools are a growing area of concern as the technology becomes increasingly prevalent in both business and consumer sectors. Here are some of the biggest fears:

- **Security issues:** AI models process incredibly large sets of data, which can hold personal information. Unless these datasets are managed effectively, they may be vulnerable to unauthorized use or manipulation. As AI platforms grow increasingly popular, the risk of breaches increases. AI systems that consume massive amounts of data, particularly in real-time, are vulnerable to attacks by hackers. They can be hacked if they do not have adequate encryption or security measures in place.

- **Data bias:** AI systems may inherit biases based on the data they are trained on to generate discriminatory or biased output. This can impact policy decisions on issues such as employment, health care, and policing, and it might harm groups or individuals.

- **Data ownership:** As enterprises use AI tools more and more, there may be concerns about who is responsible for AI data. Businesses will need to understand the ownership of the data gathered by these tools, particularly for those who use third-party AI. This has implications for responsibility, secrecy, and regulation.

- **Transparency:** Deep learning platforms are often "black boxes," and we cannot see how they make decisions. This inaccessibility can make compliance with data protection laws (e.g., The General Data Protection Regulation in the European Union) difficult to track.

Job Displacement

AI-induced job loss is the main fear of workers that comes with the widespread use of AI. With trained algorithms increasingly taking on tasks once undertaken by humans, sectors like manufacturing, retail, and customer service are particularly at risk. AI-powered automation can replace repetitive, manual labor tasks. McKinsey Global Institute estimates that between 400 and 800 million jobs around the world will potentially be lost by 2030 due to automation.[87]

Such displacement is significantly distressing because most of the affected jobs are low-skill, low-paid ones, further promoting economic inequality and social unrest. While AI has the potential to vastly increase productivity and economic growth, this gain might not be shared equally. Experts who develop or operate AI systems may be more in demand, but many middle- and lower-skill workers might lose jobs or suffer from stagnant wages. This uneven impact should alarm policymakers, as it can broaden the inequality gap and leave those who can adapt to the new economy separated from those who cannot. For instance, according to the World Economic Forum, in 2025 alone, up to 75 million jobs may be lost to AI, even as new roles may open up to 133 million.[88] Traditional industries will be more likely to lose jobs, while the healthcare and education industries are expected to see significant job growth. Dealing with job losses involves a lot of retraining and reskilling of the workforce. It is time for governments, businesses, and schools to work together to give workers the ability to survive in the AI economy.

AI may provide productivity benefits and innovation in the future, but this will not be achieved without impacting the overall structure of employment.

Copyrights Issue

Another AI-related concern is related to the issue of copyright. One of the main issues is the creation of derivative works by AI systems, which often train on large datasets containing copyrighted materials without explicit permission from the original creators. This raises questions about whether AI-generated content, such as art, music, or text, can be considered an infringement of copyright if it closely resembles or mimics existing works. A key challenge is determining who owns the rights to AI-generated creations: the developer of the AI, the user who instructs it, or the AI system itself, which lacks legal personhood. AI tools may inadvertently reproduce portions of copyrighted works, potentially violating intellectual property laws without clear accountability. The ongoing debate on this matter in the public space stresses the need for clear guidelines on AI's intellectual property rights.

Ethical Considerations

As outlined in this chapter, AI is booming and changing our world. However, according to a Workday global survey, only 29% say they are confident that AI and machine learning are being used ethically today. For project professionals, there are various ethical concerns. Naveen Goud Bobburi, Chief Manager at ICICI Bank in Hyderabad, and Lea Li, Technical Program Manager at Meta in Menlo Park, California, USA, discuss these key ethical issues in PMI's "Projectified®" podcast, referring to them as the "ABCs of morality": correctness, prejudice, obedience, discrimination, justifiability, and equality[89]. They also list several actions companies should take to protect ethics, as well as what project managers can do to stay compliant with them.

To address AI ethics, companies should create holistic AI ethics policies in which they explicitly outline their organization's position on AI ethics, as well as create guidelines around high-priority issues that AI is impacting (i.e., data privacy, fairness, transparency, accountability, safety, etc.). This action could include regular assessments of the ethical implications of projects. This could be reinforced by regular checks to identify and evaluate the potential ethical implications of projects and then to sort out any biases or privacy hazards flowing from them. Consequently, a project manager and team should be constantly trained to work with an understanding of AI ethics and to employ their best responsible AI practices.

Creating an AI Ethics Committee may be helpful for the governance of ethical discussions regarding AI. Establishing a cross-functional body to effectively assess AI issues at the organizational level, with stakeholders from all involved teams, would provide necessary support to project managers. This committee would enable PMs to collaborate with subject matter experts, such as legal professionals, and stay informed about rapidly changing regulations to ensure that project teams remain compliant.

Any issues arising from AI ethics assessments should be documented for knowledge sharing, in alignment with a growth mindset approach. A project manager with a growth mindset can lead this effort by facilitating knowledge transfer and ensuring that AI ethics training is integrated across the team and disseminated throughout the organization.

Chapter 8

Systems for Continuous Improvement in Project Management

This chapter describes the tools, frameworks, and systems that help in the continuous improvement of processes and operations used in project management. Whether it is Agile practices such as Scrum and Kanban or information-based systems like Six Sigma and Lean, these technologies offer tangible solutions to overcome problems, reduce waste, and produce scalable results. The chapter also touches on advanced practices such as Kaizen and Design Thinking, delivering creative problem-solving and cultural interventions for the creation of high-performance teams. Featuring structured tools like Value Stream Mapping, the PDCA Cycle, and the "5 Whys", this chapter gives project managers the insights they need to help their projects foster a culture of excellence and innovation.

Scrum: Driving Agile Growth Through Feedback

Scrum is an Agile framework for the collaborative, iterative management of projects. While it is usually used in the context of software development, in theory, it can be used for any task, process, or project. It helps teams manage their work through the establishment of a set of values, principles, and practices. This is the leading Agile framework used by the majority of team-level Agile users: 63% of these followed the Scrum methodology in 2023.[90] This approach to project management brings an exciting new dimension to the handling of various project management situations, such as risk, scheduling, and other factors involved in completing a project.[91]

The Scrum process uses time-boxed Sprints, typically lasting two to four weeks. At the start of a Sprint, the team holds a Sprint planning session to determine what work will be done, taking 'stories' from the prioritized product backlog. During the Sprint, the team meets daily to keep each other informed on progress and identify any impediments or obstacles to completing the work. At the end of the Sprint, there is a review to show what was done to stakeholders, and a retrospective is performed to reflect and determine what it might look like to improve the process for the next one. This rigorous approach brings the entirety of the team together around the need to deliver high-quality products in small chunks, reflecting the complexity of the work and adapting the product to the real needs of the business.

The Scrum method reinforces visibility, inspection, and adaptation within its activities—actions that could broadly be seen as the realization of growth mindset principles, reflecting the belief that skills and intelligence can be continuously improved with time and focused learning rather than being immutable qualities we are born with (or not). From time to time, it is important for everyone involved to gather as a team and reflect on their processes and Sprint outcomes to identify areas for improvement. This establishes a culture of learning and codifies the mechanism of tackling complex challenges in discrete steps, promoting resilience and agility within teams. The Scrum approach reinforces other growth mindset rules as well. Moving fast and promoting a culture of experimentation leads to teams being empowered to learn from success as well as failure. Employing Scrum promotes collaboration by ensuring cross-functional, self-organizing teams of individuals with unique skill sets who share goals and outcomes and are mutually accountable for their work.

The Scrum approach has strictly defined roles and frameworks, the key elements of which are explained in the following table.

Table 16. Key elements of Scrum

Key Elements of Scrum	
Roles	▪ **Product Owner**: Represents stakeholders, prioritizes the product backlog, and ensures the team delivers the right value. ▪ **Scrum Master**: Facilitates the Scrum process, removes obstacles, and supports the team in following Scrum principles. ▪ **Development Team**: A cross-functional group responsible for delivering product increments at the end of each Sprint.
Artifacts	▪ **Product Backlog**: A prioritized list of all desired work items. Evolves as new needs arise. ▪ **Sprint Backlog**: A selection of items from the product backlog that the team commits to completing during the current Sprint. ▪ **Increment**: The sum of all completed product backlog items during a Sprint that meets the team's definition of "complete."
Events	▪ **Sprint Planning**: A meeting where the team determines which items from the product backlog will be worked on in the upcoming Sprint. ▪ **Daily Scrum**: A short (15-minute on average) daily stand-up meeting to discuss progress and challenges. ▪ **Sprint Review**: A meeting held at the end of the Sprint to showcase the completed work to stakeholders and gather feedback. ▪ **Sprint Retrospective**: A meeting to reflect on the Sprint, discussing what went well, what did not, and how the team can improve.
Sprints	▪ **Time-Boxed Iterations** (typically two weeks) focused on delivering a product increment. Each Sprint results in a potentially shippable product piece, allowing for frequent delivery and iterative development. ▪ **Work Progress** is measured in velocity, showing how many user stories have been completed.

Scrum framework offers multiple benefits for the project teams:

- **Faster time to market:** This iterative approach to development allows teams to deliver results more quickly than they would using the traditional Waterfall methodology.
- **More collaboration:** By encouraging teamwork and communication, Scrum contributes to better collaboration among team members.
- **Incremented quality through continuous feedback:** Increasing the frequency of feedback loops helps identify early issues, thereby improving the quality of the product.
- **Improved responsiveness to change:** Scrum helps teams adapt their plans and priorities more easily as they evolve.
- **Develop a growth mindset:** Scrum encourages team members to look at obstacles as challenges and opportunities for improvement.
- **Growing resilience and innovation:** A growth mindset leads to resilience, innovation, continuous learning, and improvement.
- **Authority to experiment:** Teams are given the authority and encouragement to experiment (with room for failure as a learning opportunity).
- **Freedom to fail & try new things:** Everyone on the team is encouraged and excited about taking risks and trying new things, bolstered by an understanding that not everyone will be successful in every project.
- **Boost in motivation and engagement:** The concentration on both personal and team flourishing directly helps to upgrade each peer's individual abilities, reflecting a high level of group output.
- **Dynamic and flexible work environment:** By empowering team builders with a growth mindset, organizations can instill such an environment through Scrum.
- **Positioned for sustained success:** The Scrum approach prepares organizations to succeed in a dynamic environment and fosters lasting resilience.

The Role of Feedback in Scrum

Feedback events are a key part of a Scrum, mainly because they promote collaboration and ensure communication is as free-flowing as possible.

- The **Sprint Review** is held at the end of a Sprint to allow members from product management, customers, prospective users, or other stakeholders to see what functionality and progress has been achieved by the development team during that time increment. During the event, the team demonstrates their work from those two weeks and celebrates its successes, as well as discusses any obstacles or challenges that were faced in order to gather more productive feedback.
- The **Sprint Retrospective** is held immediately following the Sprint Review and allows the Scrum team to review strategy with a focus on continuous improvement. This event addresses what worked, what didn't work, and how things could be improved–resulting in actionable outcomes for future Sprint improvement.
- For large projects with more than one Scrum team, the Scrum methodology uses the **Scrum of Scrums** mechanism, where representatives from all teams come together to discuss progress and interdependencies and address any arising frictions. These events cultivate a culture of openness and transparency, allowing teams to change continuously and iterate processes for improvement.

Lean: Optimizing Processes for Continuous Improvement

The Lean methodology offers project managers a robust set of tools, allowing them to systematically identify and improve bottlenecks and inefficiencies at various stages of the project life cycle, as well as help build scalable teams via improved resource allocation and resource utilization. This will lead to better project outcomes. Visualization methods enable quick and prompt corrective action to ensure the project remains on time, within the budget, and of high quality, delivering goods or services that surpass client expectations.

Additionally, the Lean methodology boosts project teamwork and overall cross-functional collaboration by constantly encouraging members to communicate openly and constructively. It creates a growth mindset culture of mutual respect and support, with a general understanding that the goal is to find solutions and not to judge or criticize.

Lean originated in the early 20th century at the Ford Motor Company, where Henry Ford pioneered the concept of waste reduction and continuous production flow through his assembly line methods, which laid the foundation for what is now known as Lean. Later, after WWII, Lean was adapted by Toyota Motor Corporation in Japan and developed into the Toyota Production System (TPS)—a management philosophy aimed at maximizing value while minimizing waste. In the face of labor shortages and pressure to speed up production and boost product quality, Toyota found ways to reduce costly inefficiencies. Just-in-time production—where only the required component is produced—and worker-controlled production cessation by all the operators in a given assembly area are the two core principles of the TPS and, ultimately, of Lean itself.

In 1996, in their book *Lean Thinking*, the authors James Womack and Daniel Jones extended the system to the world of services and goods outside manufacturing. They argued that the TPS could be applied anywhere to deliver superior efficiency by emphasizing common principles and practices across systems. The fact that so many different types of industries have implemented Lean serves as proof of this adaptability they advocated for. Hospitals and clinics streamline healthcare processes to reduce wait times and improve care delivery. In software development, frameworks such as Lean Software Development can be employed to deliver customer value more quickly and effectively. In service industries such as banking or hospitality, Lean is used to improve operations and add value for the customer. Startups, which are still establishing processes, and large enterprises that need to stay ahead of the competition both use Lean methodologies to maximize innovation and optimize performance.

Lean principles promote the development of a growth mindset among teams and organizations. Emphasizing the necessity of continued improvement and the opportunities for learning from experiences of failure shifts the way in which challenges are seen by those involved in implementing the Lean methodology. Instead of being expected to see failure as a discouraging sign of incompetence, individuals who are well-versed in the Lean approach are encouraged to view challenges as opportunities to learn and improve. When actively employing this approach, Lean teams (or organizations) are likely to become more resilient and flexible in their management style.

Lean uses various techniques to help achieve the broader goal of minimizing waste and maximizing value. Some of these techniques support project management activities, providing a deeper understanding of processes, risks, and opportunities.

Value Stream Mapping: Efficiency and Waste Reduction

Value Stream Mapping technique consists of creating diagrams illustrating the comprehensive material and information flows required by an organization to create a product or provide a service to customers. Mapping out the process helps identify redundancies, and waste.

These maps start with the customer, putting them in the center of all Lean operations, and teams look at current as well as future Value Stream Maps to see where they can make improvements, eliminate waste, and deliver a product faster. By mapping out every step of the process and identifying what adds or reduces value (inside the circle or outside the circle, respectively), inefficiencies in the process become easier to spot, thereby facilitating the creation of strategies to reduce them as well as to shorten the time it takes a product to get to the customer. A Lean practice, for example, might involve placing machines next to the worker instead of the other way around to save time and reduce costs.The diagram below illustrates a Value Stream Mapping example for software development, presenting lead time and value-added time for each project phase.

- **Lead Time:** The total time it takes to complete each stage, including delays or waiting periods.
- **Value-Added Time:** The time spent directly contributing to customer value at each stage.

Figure 11. Value Streaming Mapping Chart

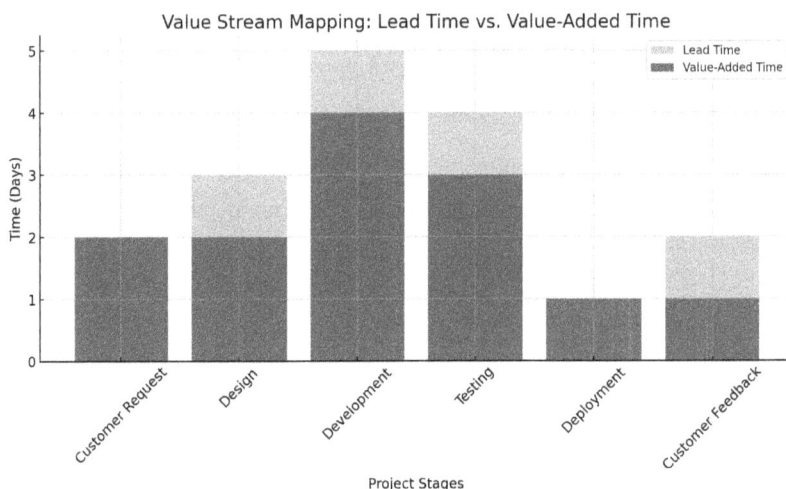

From the initial Customer Request to the Customer Feedback phase, a breakdown of the total lead time versus value-added work time highlights areas for potential enhancements, which can be addressed to streamline the overall delivery process.

Kanban: Visualizing Workflow for Efficiency

Kanban is a method for managing projects that help teams see what they're working on, plan together, and continuously improve. It is a pull-type system that allows work to start where and when it's needed, consistent with Lean principles. This methodology limits work-in-progress to a predetermined capacity, balancing team demand against the team's throughput of delivered work. This allows teams to find a sustainable development pace so that individuals achieve a sustainable work-life balance. Kanban plays a pivotal role in enhancing quality and performance by controlling amounts of work-in-progress. It reduces lead times, improves predictability, and ensures on-time performance, thereby demonstrating its effectiveness in multiple project management efforts.

Below is an example of a task tracker designed according to Kanban principles. It provides an overview of the workload to be done, in progress, under review, and completed.

Figure 12. Example of a Kanban Board

TO DO	IN PROGRESS	IN REVIEWS	DONE
Conduct user interviews.	Draft problem statement.	Design prototypes for users' research.	Analyze survey results.
Brainstorm solutions.	Schedule user review sessions.		Publish insights.

Utilizing Kanban improves the quality of a team's work and performance. It is highly relevant to organizational culture because it's a problem-solving, problem-preventing, and solution-focused tool that helps to move an organization forward. Within this framework, individuals are encouraged to cooperate and have complete visibility of the project team workload.

Kanban helps create complex, adaptive, and emergent behaviors in an organization. To do so, Kanban uses five core properties that have been successfully implemented in top global companies, including, e.g., Microsoft:[92]

- Visualize Workflow
- Limit Work-in-Progress
- Measure and Manage Flow
- Make Process Policies Explicit
- Use Models to Recognize Improvement Opportunities

The benefits of Kanban include greater customer satisfaction, frequent delivery of value, increased productivity, improved quality, and improved efficiency. Most importantly, it helps the organization become agile through cultural change.

5S for Waste Reduction

5S is a structured methodology designed to eliminate waste and maximize efficiencies by maintaining a clean and organized workplace and through the use of visual techniques to optimize consistent operational outcomes. When applied, worksites are purged of clutter, rearranged for optimal functionality, and configured in accordance with what has already been determined.

The pillars of 5S are Sort (*Seiri*), Set in Order (*Seiton*), Shine (*Seiso*), Standardize (*Seiketsu*), and Sustain (*Shitsuke*).[93] This framework facilitates a logical and sequential way of structuring a clean, organized, and effective work environment. In operational terms, performing work according to established routines that maintain organizational structure and order encourages a clean and organized work environment, allowing for a more seamless and effective workflow. This method enables employees to enhance their workplace and assists them in learning how to minimize waste, unplanned downtime, and the work-in-progress inventory.

Some of the world's largest companies have successfully used the 5S technique to help improve their operations. For example, Boeing has used the 5S technique to improve processes and safety, as well as to reduce clutter and injuries in the company's manufacturing plants. Boeing has also created cleaner and more conducive work environments that have led to more productivity and less assembly time. 3M has similarly used 5S to improve manufacturing and research efficiencies in its factories and research labs, facilitating a more creative and productive work environment. The emphasis on cleanliness and organization in these environments has also helped to improve the company's product development processes. Nestlé, one of the world's largest food and drink companies, also uses this system to ensure that its manufacturing plants are kept clean and organized. Since food is the company's main product, keeping these plants clean and ordered is fundamental to ensuring high-quality standards in food production.

All of these companies represent how the 5S methodology can help to improve operations across different business types and operations.

Just-in-Time (JIT) for Adaptive Project Delivery

Lean Just-in-Time (JIT) production involves a manufacturing approach that reduces waste and maximizes efficiency by producing only what is needed, when it is needed, and in the required quantity. It maintains inventories at a minimum by matching production quotas to actual demand and minimizing overproduction to lower storage costs. Its components include a continuous production flow, small batch sizes, integrated supplier coordination, and waste reduction. By ensuring materials and products arrive 'just in time' for use, JIT production increases responsiveness, lowers costs, and heightens efficiency; however, in this context, demand forecasting and reliable suppliers are crucial.

Just-in-Time production, developed in the manufacturing industry, has been successfully translated into other sectors. In the tech industry, for example, particularly in software development, it supports smaller and more frequent iterations of coding output by providing alternative ways to fix bugs and minor UI issues. Instead of accumulating elements for a more significant change, JIT helps to reduce waste by delivering 'just-in-time' updates only as they are needed, preventing the overproduction of features that are either not used or could quickly become obsolete. This approach works synergistically with Agile development processes and Continuous Integration/Continuous Deployment pipelines, which enable teams to push small, incremental updates frequently, thus catching the bugs earlier and allowing for smoother deployments and higher overall code quality. Real-time feedback from users also helps with immediate adjustments. It promotes the cycle of Kaizen (continuous quality improvements), which is crucial for maintaining a competitive advantage in fast-paced marketplaces.

JIT benefits in the tech industry may be substantial: increasing time to market by getting features into production faster, allowing teams flexibility and the ability to pivot based on user feedback and changing market trends, keeping individuals and teams responsive to evolving environments, enabling the efficient use of resources, and reducing risks.

Applying Lean for Scalable Project Management

Adapting Lean as a corporate-wide philosophy may be challenging. Developing a systematic and holistic way of thinking about the business starts with a thorough situational analysis to address the following critical strategic questions: "What is our business model, and what are the needs regarding business operations?"; "How do we make money?"; "Is Lean the right fit for the organization?"; "Is there another theoretical business model that might be a better fit for us?"

Although Lean might not always be the right fit for every aspect of the business, it is an excellent way to develop a culture of continuous improvement through a growth mindset by tackling waste, improving processes, and encouraging teams to suggest and resolve inefficiencies. Organizations can foster an environment in which every employee believes they can contribute ideas that improve how they work every day.

For project managers specifically, the principles and processes of Lean are helpful in small yet meaningful ways. Even if scaling Lean across the organization is beyond your authority, you can commence by piloting Lean processes on your projects or in your department. This small step can significantly impact the company's progression towards greater efficiency and agility.

The following is an example that demonstrates how Lean can facilitate the redesign of a single business process, making it safer, faster, easier, and more reliable. Lean's focus on the concept of removing inefficiencies, which will be explained below, can lead to significant performance improvements in a narrowly defined area, with the potential to spread to other areas of the organizational system over time.

Lean Application: Example

Harsha, a project manager in the Go-To-Market team of a large retail enterprise, was tasked with improving the delivery cycle of market research projects. Before changing the main project management framework, the average delivery time for a typical market research project was 12 weeks. The goal was to reduce this by at least 30%, to a maximum of 8-9 weeks, while maintaining the same deliverable quality.

After identifying delays in data analysis and reporting, he brought in the Lean methodology to streamline the process. In the initial project trial, Harsha implemented the following:

- A process map, leveraging Value Streaming Mapping to assess what led to the greatest inefficiency, which he then eliminated through collaboration with senior management.
- A new cross-functional collaboration space (Confluence) where researchers, analyst and stakeholders could come together.
- Trimmed down the number of meetings, concentrating only on those that were essential.
- Created a Kanban board to monitor daily progress and maintain team motivation.

Results:

- The goal of reducing the project delivery time was successfully achieved, with the new turnaround time set at 8 weeks.
- The teams experienced reduced stress levels and increased productivity.
- Enhanced research quality and faster insights were achieved.

The pilot's success led to the adoption of Lean practices throughout the retail department and to a radical change in how projects were delivered across the corporation.

Kaizen: A Mindset for Continuous Improvement

Kaizen is a management strategy based on principles that encourage a growth mindset. The concept refers to business activities that are aimed at constant and incremental improvement of all functions in a company. Its critical principles are as follows:[94]

- **Standardization and adaptability:** By establishing, documenting, and sharing best practices and lessons learned as a foundation for incremental and sustainable improvement, organizations can more easily pinpoint deviations, challenges, and new areas for growth while maintaining flexibility.
- **Gemba focus ('shopfloor' emphasis):** Through directly observing work processes, interacting with employees, and addressing issues collaboratively at various levels of an organization, leaders aim to continuously enhance productivity, quality, safety, and morale in a manner respectful of diverse perspectives.
- **Continuous improvement:** Promote flow efficiency by methodically eliminating inefficiencies and waste, such as unnecessary costs, defects, and rework, while streamlining and adapting operations in response to changing needs and opportunities.
- **Customer focus:** Optimize value creation by accurately understanding and proactively serving the needs of both internal and external customers to build loyal relationships and a solid reputation.
- **People respect:** Individuals are the core of improvement initiatives, fostering an environment built on mutual trust and empowerment and showing regard for employees who actively contribute to identifying and implementing enhancements through dedicated effort and teamwork.

Kaizen originated in Japan; however, it was influenced by the American methodology of Total Quality Management, which was implemented, for example, in Ford factories. Kaizen is a combination of two Japanese words: "kai" (change) and "zen" (better), literally meaning "change for the better." It is often translated into English as "continuous improvement."[95]

Kaizen, which we know today as a comprehensive methodology, was created primarily by the Toyota company. The "Toyoda Precepts," established by Sakichi Toyoda, the founder of Toyota, and his sons Kiichiro and Risaburo, represent the foundational principles that molded Toyota's legacy. Prior to entering the automotive sector, the company was renowned for generating pioneering weaving machines, a business that laid the groundwork for its eventual triumph. Sakichi Toyoda's expertise in loom-making, coupled with Kiichiro's visionary approach, led to the development of self-operating looms that found widespread success. The guidelines steering Toyota's evolution were later codified as the Toyoda Precepts, emphasizing team effort, innovation, usefulness, kindness, and gratitude.[96]

Nowadays, continuous improvement is a fundamental aspect of Kaizen's operational excellence, focusing on continually enhancing an organization's processes, services, and products. It promotes a proactive stance towards problem-solving and process refinement, highlighting the presence of opportunities for improvement despite the efficiency of current systems. The objective is to cultivate an environment of enduring excellence instead of solely tackling isolated concerns.

The Five Elements of a Kaizen Culture

Kaizen consists of five founding elements, which are the spine of the entire methodology:[97]

- **Teamwork:** All employees should work together to achieve one end goal, which is to continuously improve internal business processes. In other words, "all employees" refers to anybody from corporate to front-line employees.

- **Personal discipline:** Self-disciplined employees make better and more productive members of an enterprise. Therefore, Kaizen philosophy demands that all employees increase their self-discipline in all aspects pertaining to their work—the way they manage their time, their performance, their expenditure of material and financial resources, how they collaborate and so on.

- **Improve morale:** A company may or may not achieve these changes in its workforce. However, all workers should do their best to keep their motivation high and increase their confidence and enthusiasm. To facilitate this, senior employees should help their workers with specific motivation tools: for example, by providing pleasant working conditions, incentives and rewards, and good salaries and benefits.

- **Quality circles:** Circles that include staff at different degrees of the business. Circles of quality mean that staff are given the chance to discuss and create ways to improve internal business processes. These groups are able to share ideas, knowledge, technology, and other tools for staff to use. This encourages employees to work together and cooperate more effectively. The staff can assess processes' efficiency and find ways to improve them further.

- **Suggestions for improvement**: All employees should have the right to freely express their ideas and suggestions to enhance the running processes. Moreover, if it is feasible, suggested ideas should be immediately considered and implemented if deemed useful.

Although rooted in the idea of continuous improvement, one of Kaizen's important themes is the fight against perfectionism, which we can interpret as encouraging an evolutionary view of change in which assumptions are challenged rather than taken for granted.

The similarities between these features of the growth mindset and Kaizen management, as highlighted above, suggest a shared emphasis on collaboration, adaptability, and lifelong learning.

With the rise of Japanese multinational companies and the boom of globalized production, the international transfer of Kaizen has become very important. Since most production companies are in search of higher quality and more cost-effective products, the adoption of Kaizen has a huge market potential globally. However, it is worth noting that the success of its transfer varies from one country to another, and there are a few factors that influence the possibility of Kaizen succeeding overseas. Proper preparation in the Kaizen mindset requires employee commitment, open communication, skills development, focus on continuous improvement, and team building. Although some researchers hold the view that Kaizen can be adjusted to use outside Japan, some believe that it is challenging to do so because of its deep-rooted cultural characteristics.[98]

Some organizations might consider adopting Kaizen fully, while others may need to implement significant organizational changes to foster such a growth mindset environment. Nevertheless, certain selected Kaizen techniques are valuable for investigating and applying in project management practices.

5 Whys Technique for Deep Problem Solving

The questioning technique known as the 5 Whys method is similarly attributed to Sakichi Toyoda, founder of the Toyota company, and to the architect of the Toyota Production System, Taiichi Ohno. This method involves iteratively asking "why" several times, using each answer to form the next question, and progressively getting closer to the root cause of a specific problem.[99] The fundamental principle underlying these methodologies is to engage the human mind in a structured process of inquiry, using a series of questions to promote problem-focused thinking.

Below are key points emphasizing the significance of the 5 Whys Technique:

- **Problem-solving depth**: By delving beyond surface-level reasoning, the 5 Whys method prompts individuals to unearth the fundamental cause of a dilemma layer by layer. It aids in dismantling intricate issues into more manageable segments that can be addressed effectively. Each discovery paves the way for new insights and solutions.
- **Learning opportunities:** Each subsequent question elicits a fresh perspective and deeper comprehension. This progressive approach cultivates a growth mindset and culture of continuous learning and advancement within organizations. Employees are encouraged to challenge assumptions and think critically.
- **Collaboration and communication:** Applying the 5 Whys often involves cross-functional teams, rendering it a collaborative endeavor. This joint effort introduces diverse viewpoints and expertise that might otherwise go unnoticed. The questioning spurs rich discussions that strengthen relationships and build shared understanding.
- **Preventive action:** Uncovering the root cause enables targeting solutions at systematic flaws rather than symptoms. This empowers teams not only to remedy current problems but also to introduce proactive measures for the future, always in harmony with continuous improvement principles. Issues are resolved at their source.
- **Openness and curiosity**: Effectively employing the 5 Whys technique demands the resistance of easy answers in favor of exploring each layer of complexity. It transforms mindsets from fault-finding to investigating with patience and perpetual growth in mind. Teams remain open to new lessons and committed to excellence.

The 5 Whys technique cultivates a growth mindset—it encourages a questioning attitude and aims to move forward, anticipate, and take things to the next level. How does it work in practice? Below is an example.

The 5 Why Technique at Work: Example

SYNTH, a fictional software company, delivers business services to its clients on a subscription basis. The majority of the customers make payments on monthly plans.

Usually, a significant portion of the payments is made on time. However, this month, the company has noticed that about 25% of the payments are behind schedule. That is an oddly high number for a normally well-functioning organization. This surely calls for a prompt intervention from the team.

A project manager named Jason is responsible for investigating this issue. He decides to form a cross-functional team composed of Product, Business, Engineering, and Finance groups. He chooses the 5 Whys framework to analyze the situation. He begins by defining the problem and asking the right question. After he investigates the problem, he proposes the steps listed below.

5 Why Framework Used by SYNTH

Define the problem: This month, the software company SYNTH observed that 25% of payments for service subscriptions from customers is delayed. Delays increased by 90%, compared to last month.

1st Why:

- Why are customers' payments delayed 90% more than last month?
- Answer: Because many customers have not made payments on time.

2nd Why

- Why have many customers not made payments on time?
- Answer: Because they are encountering issues or difficulties in the payment process.

3rd Why

- Why are customers encountering issues or difficulties in the payment process?
- Answer: Because the online payment system has had technical issues in the past few weeks.

4th Why

- Why has the online payment system had technical issues?
- Answer: Since the introduction of the newest system update, SYNTH has been experiencing problems where customers who try to make a payment are confronted with errors.

5th Why

- Why did the latest system update have errors?
- Answer: As a result of the update being rushed out without proper quality assurance, so as to meet internal deadlines.

Define root cause: The main reason seems to be that the payment system update was not adequately tested and has bugs. This has caused technical problems that are holding up payments to customers.

Fix the issue: Investigate and resolve the current technical issues as quickly as possible. Consider setting up a temporary alternative payment method until the system is stabilized.

Improve process: Implement a robust testing process for system updates to avoid future bugs.

Jason and the team identified the issue causing customers to delay payments: a bug in their information system and a gap in the quality assurance process. Recognizing this setback as an opportunity for improvement, they adopted a growth mindset, focusing on refining their processes and enhancing their system to better serve their customers moving forward.

PDCA Cycle: Iterative Growth in Action

The PDCA (Plan-Do-Check-Act) cycle serves as a fundamental framework for driving continuous improvement efforts within organizations. It offers a cyclical approach that facilitates systematic planning, testing, evaluation, and optimization of changes.

Through enabling organizations to experiment with enhancements on a small scale and make adjustments based on data and feedback, the PDCA cycle encourages the sustainable enhancement of processes, products, and services, ultimately contributing to operational excellence. The PDCA stages are as follows:[100]

- **Plan**: During this stage, the focus is on identifying opportunities for improvement and creating a detailed action plan to tackle them. This entails establishing specific goals, choosing key performance indicators for success, and forming hypotheses on how the proposed changes can lead to enhancements.
- **Do**: At this point, the planned modifications are put into action on a small scale or within a controlled setting. This phase allows organizations to test the feasibility of the proposed solutions without causing significant disruption to daily operations.
- **Check**: Following the implementation, the outcomes are meticulously examined and compared against the goals defined in the planning phase. This step is vital for evaluating the effectiveness of the actions taken and pinpointing any deviations or areas needing enhancement.
- **Act**: Drawing on the insights and data gathered, organizations determine whether to roll out the changes on a larger scale, make adjustments, or discard them. If the results are positive, the enhancements become standardized and adopted as new practices. Otherwise, the process begins anew with fresh hypotheses and strategies.

Figure 13. Plan-Do-Check-Act Cycle

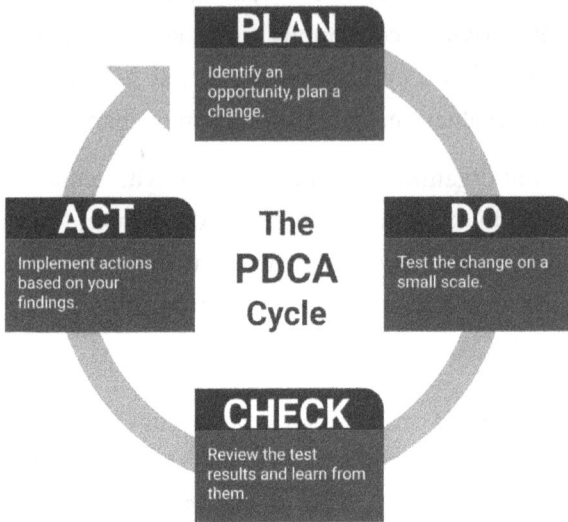

Standardization for Scalable Growth

Documented best practices are among the most effective (and underappreciated) tools within Kaizen and Lean. Once a standard has been developed, capturing the very best method for getting work done, it creates a foundation for a cycle of continuous improvement. Over time, that standard can evolve into a new best practice.

Standardizing work is a key factor in supporting continuous improvement within any organization. It helps to accelerate, optimize, and reduce costs and timelines. The business insights gained from routinizing work processes immediately improve efficiencies and performance, positively impacting the bottom line. It also sets a measurable standard for evaluations and assures quality performance, which can be the foundation for automation and process innovation, further supporting ongoing growth and development.

The process of standardization—which involves a growth mindset— opens the door to endless improvement and learning. It means an organization can give teams the stability to play with variations on the standard, experiment, innovate, and adapt without compromising quality. Moreover, it teaches people not to be deterred by failure because failure becomes simply a framework to measure further effort, from which the gratification of improving the projects will arise.

Standardization in Practice: Example

Flow, a fictional software company, provides consultancy services to the banking sector. Managers found significant inefficiencies due to variances in project management practices, and they thought they could improve outcomes by standardizing and enhancing processes. Mark, a program manager in Flow's PMO, has been asked to identify pain points caused by variances and to define standards to remove inefficiencies and improve teamwork. The following outlines his actions.

Before standardization

- **Inconsistent methods:** Project managers used different tools to track tasks and teams managed risks inconsistently, which created confusion and inefficiencies.
- **Poor communication:** The status of projects was not updated frequently and regularly, leading to misunderstandings between stakeholders.
- **Project delays:** The product owner increased the scope after the project kick-off, causing team members to take too long to finish tasks. The product owner and project manager used different ways of planning milestones, and there was no single way to identify constraints before starting projects.

Pain points addressed by the project manager

- **Better visibility:** New reporting and tracking systems were needed to make a project's progress more visible, allowing management to spot bottlenecks quickly.
- **Managing risks:** The company needed to improve its risk-resolving methods and increase its risk resolution speed.

- **Improved collaboration:** Without standardized tools and reporting requirements, team communication broke down, causing misunderstanding and delays.

Process standardization and improvement solutions

- **Task & resources tracking:** A single project management tool (Jira) was mandated to ensure a standardized workflow across all teams. The tool was configured to send automated notifications and track assignments and dependencies, creating a more fluid flow of tasks across projects.
- **Risk management:** Common risk assessments were standardized through the implementation of a risk register, with all teams assessing risks using the same categories. In addition, risk monitoring features were established to provide early warning signals for potential bugs or delays.
- **Communication:** Status reports are standardized using the same template and are expected to be brief, including key performance indicators (KPIs) such as the percentage of tasks completed, risks associated with the project, and deadlines, making it easier for the stakeholder to understand the status of the project and the upcoming challenges.

This standardization led to greater efficiency in Flow because it clarified to the team what they needed to do and helped them develop an 'ongoing spirit of growth' as a company. The team has become more transparent and in sync with each other, allowing them to focus on 'continuous improvement' and 'joint effort.'

Six Sigma: Data-Driven Excellence

For decades, Six Sigma has promoted measures of variability in corporate processes to gradually reduce defects on the assembly line. By more reliably meeting customer needs while also reducing friction in manufacturing workflows, the methodology—developed at Motorola in the 1980s—is meant to eliminate waste and maximize efficiency. Its emphasis on total quality through scientific methods draws on 19th-century ideas about control in industry. However, Six Sigma is no longer a proprietary protocol of any particular brand; today, its statistical methodologies dictate improvements at global giants such as Amazon, Boeing, Ford, and Samsung. Even the US Army, Navy, Marines, and Air Force have adopted Six Sigma to sharpen performance across their disparate operations.[101]

Six Sigma, usually represented by the notation 6σ, focuses on using data to make decisions. Measurements and metrics are leveraged to determine the quality of a company's operation, the cost of manufacturing goods or services, and the time it takes to complete them. Six Sigma is designed to improve the quality and efficiency of the way things are done. At its heart, Six Sigma uses rigorous statistical analysis of processes—inputs and outputs—to eliminate defects.

Key metrics of the approach include Defects per Million Opportunities (DPMO), which provides a comprehensive analysis of process performance by precisely quantifying defects on a per-opportunity basis; Rolled Throughput Yield (RTY), which measures the likelihood of producing flawless items while accounting for rework serving as a holistic gauge; and Process Capability, which assesses the extent to which a process can fulfill product or service specifications, minimizing unacceptable variance. Additionally, other important considerations include Overall Equipment Effectiveness (OEE), encapsulating process proficiency with regard to scheduling and defects, and Customer Satisfaction, which is paramount in Six Sigma due to its emphasis on the effect that business decisions have on clients. Moreover, traditional measures like Yield and Throughput Yield still function as complete evaluations of process output.[102]

The Six Sigma approach provides a holistic view, encompassing suppliers and all the inputs they provide, internal processes, outputs, and customers.[103] Managers in diverse industries have adopted Six Sigma since it provides a clear rationale for the value of their investment, galvanizes employees around targets, and instills accountability across the organization, all while synchronizing various metrics companywide. Just as importantly, it centers the customer at the core.

Although Six Sigma is based on the same premise as a growth mindset— that every talent can be cultivated through effort, practice, and study— a growth mindset can work only if it is matched with practice. Good teams using Six Sigma relentlessly pursue perfection, subject their results to rigorous analysis, and quickly translate the learnings drawn from experiments—including the frustrating ones that end in obvious failure—into action. Each iteration involves tightening up processes and ironing out misunderstandings.

Moreover, thinking about problems as opportunities to develop rigorous evidence rather than obstacles to be sidestepped builds a culture of curiosity in which obstacles are welcomed by team members and vigorously attacked. The identification of previously obscured root causes through data-guided investigations leads to the more accurate targeting of inefficiencies. Refined in the fire of continuous improvement, operations emerge with newfound elegance, speed, and clarity of purpose.

Through encouraging staff to participate in this process from the beginning stages through to completion, Six Sigma helps companies develop a strong sense of commitment and responsibility. This ultimately produces engaged, well-trained workforces, which demonstrates the value of doing business together through cooperation. In the same vein, collaboration is also key in this context. Handling difficult tasks with customers, suppliers of services, or suppliers themselves involves sharing different points of view and working collectively to overcome obstacles to achievement.

Problems that are confronted with a growth mindset and the Six Sigma approach are dealt with constructively, leaving no legitimate reason for anybody to give up just because the first attempt failed. By providing resources for structured learning programs like Belts (White, Yellow, Green, and the most advanced black Belt) and Certified Six Sigma training, this philosophy not only develops people with the needed technical skills but also supports their continuous development.

Frameworks for Data-Driven Growth

Six Sigma utilizes two methods for process improvement: DMAIC and DMADV. The choice of method depends on whether the goal is to improve existing processes or design new ones.

DMAIC, which stands for Define, Measure, Analyze, Improve, and Control, is employed to enhance processes that already exist. It begins by clearly articulating the problem and the goals the team wants to achieve, followed by quantifying how the process currently performs, investigating the root causes through analysis of collected data, carrying out modifications to fix issues, and putting controls in place to sustain the changes.

Table 17. DMAIC Framework

DMAIC	
Framework based on a data-driven approach used to improve existing processes and designs, providing a structured way to manage controlled change.	
Define	Define the problem and project goals.
Measure	Measure key aspects of the current process in detail.
Analyze	Analyze data to identify the primary flaw in the process.
Improve	Improve the existing process based on findings.
Control	Control future implementation of the improved process.

Alternatively, DMADV, which is short for Define, Measure, Analyze, Design, and Verify, is applied when forging new processes or developing products. This approach launches by establishing objectives and understanding customer demands, gauging pivotal factors, exploring alternative designs, crafting a solution, and validating its effectiveness through experimentation.

Table 18. DMADV Framework

DMADV	
Framework aimed at developing a new process, product, or service when existing processes cannot be improved enough to meet customer needs.	
Define	Define the purpose of the new project, product, or service.
Measure	Measure essential components of process and product capabilities.
Analyze	Analyze data to explore design alternatives and select the best option.
Design	Design the chosen alternative and test a prototype.
Verify	Verify the design's effectiveness through simulations and a pilot program.

While both strategies aim to refine processes, DMAIC centers on optimizing current systems, whereas DMADV is utilized to engineer new ones that satisfy evolving customer needs. Complex processes demand balancing uniform procedures with occasional deviations; a metronomic rhythm risks monotony, whereas irregular pulses invite inconsistency.

Six Sigma Process Improvements at Work: Example

Rexton4, a fictional software service provider developing tools for digital creators, often misses project release deadlines, causing delays and customer disappointment. Xiu, a lead project manager, was assigned to investigate this issue and propose an improvement plan. She chose the DMAIC framework to complete this task. Below is a summary of her actions.

DEFINE

- **Objective:** Decrease project delays by 20% in the following six months.
- **Impacted stakeholders:** Project leads, developers, QA team, and business analysts.

MEASURE

- **Process Metrics:** Data on project timelines and delays.
- **Baseline:** 60% of projects are delayed by 15% beyond their estimated finish dates.

ANALYZE

- **Root Cause:** Frequent delays in obtaining client requirements.
- **Reason:** (1) Poor communication between developers and testers leads to rework; (2) Inconsistent resource availability across teams.

IMPROVE

- **Solutions:** (1) Hold more regular meetings with clients to clarify requirements upfront; (2) Set up daily team check-ins between developers and testers to identify issues early; (3) Enhance resource planning across projects; (4) Automate testing to accelerate the process and reduce errors.

CONTROL

- **Monitor Improvements:** (1) Track on-time delivery rates and project progress; (2) Periodically review resource allocation and testing processes; (3) Collect client feedback to ensure expectations are met; (4) Record new procedures to keep improvements consistent.

OUTCOME

- Six months later, Rexton4 reduced project delays by 25%, and the organization reported smoother collaboration between teams, leading to better on-time delivery and increased client satisfaction.

This simplified DMAIC process helped the company keep projects on track by solving point problems and staying focused on improvement efforts over time.

Six Sigma's Role in Growth Mindset Organizations

Unlike traditional quality programs, where process improvements might be sought and, once achieved, the program might be considered complete—only to develop a new goal or reset the quality objective—Six Sigma enables continuous process improvement, allowing both 'small' and 'large' changes to be implemented to improve efficiencies and reduce costs.

The principles of Six Sigma are designed to harness complexity through structured processes of continuous improvement. By breaking problems into comprehensible pieces, teams can independently study each aspect and gain a holistic understanding. This decomposes complexity through data-driven inquiry, reducing variation, measuring performance, and focusing on productivity. Whether refining existing workflows with DMAIC or pioneering innovations with DMADV, Six Sigma's methodical approach challenges complexity.

Cultural changes cultivate an innovative spirit as well. Nurturing curiosity and collaboration fosters a learning mindset that is comfortable with progress. Employees are given meaningful responsibilities through empowerment and train one another via upskilling. Teams tackle obstacles together, encouraging growth. Well-defined, quantifiable goals inspire achievement without stifling autonomy. By harmonizing methodology and flexible thinking, Six Sigma transforms associations through a culture of mutual support. Shared accountability promotes the development of both processes and people at every level. Continuous, interdependent progress fosters identification and energizes communities. This virtuous cycle maximizes potential through structured self-improvement, flexibility, and mindful adaptation to change. Together, discipline and adaptability optimize outcomes and nurture an innovative organization.

In *The Six Sigma Philosophy*, the authors Antonieta Lima and Carlos Riesenberger list the elements of this method:[104]

- **Teamwork**: Six Sigma facilitates the generation of ideas, promotes cooperation, and fosters mutual support among employees to achieve common goals.
- **Support**: Cooperation powers dedicated labor and bold achievements. Six Sigma emphasizes the importance of customer satisfaction as a means of heightening commercial importance, with a firm concentration on both internal and external customers.
- **Stimulation**: Six Sigma inspires groups by providing an organized troubleshooting methodology. An energized person can tackle new jobs, understand new techniques, or launch their own advancement campaigns.
- **Resourcefulness**: Efficiency in Six Sigma means using internal and external resources in the best way to achieve results. It encourages studying the inputs and outputs of a process to optimize performance.
- **Balance**: Six Sigma teaches a balancing of daily duties and exciting projects. This balance sparks creativity and allows workers to try new approaches to doing what they do best—making stuff up as they go along.
- **Ownership**: Six Sigma affords individuals value at all levels of the organization. It is not only the CEO who is responsible for improvements but also the switchboard operators and every other employee in an organization. Everyone brings their best abilities, skills, and experiences to maximize production.
- **Focus**: Six Sigma targets the key variables that affect processes, focusing with precision on what needs improvement.
- **Leadership**: In Six Sigma, leading by example is essential. People at the management level must set the standard for improvement, and from top to bottom, all employees should take an active part in ensuring processes run smoothly.
- **Learning from mistakes**: Six Sigma encourages learning from failures and turning errors into opportunities for improvement. An error becomes a mistake only when a person refuses to correct it.

Design Thinking: Innovating with Creative Problem-Solving

Design thinking is a human-centered, iterative approach to creative problem-solving. It puts users' needs at the core of the design process and allows organizations to experiment with solutions until they find something that 'works' for their end-users. In doing so, design thinking contrasts with traditional approaches to problem-solving, which often rely on 'analytical' thinking and on linear sequences of planning, implementation, and evaluation. This approach encourages collaboration across disciplines, enabling diverse perspectives to shape the final solution.

The five phases of design thinking process are illustrated in the diagram below:

Figure 14. Design Thinking Phases

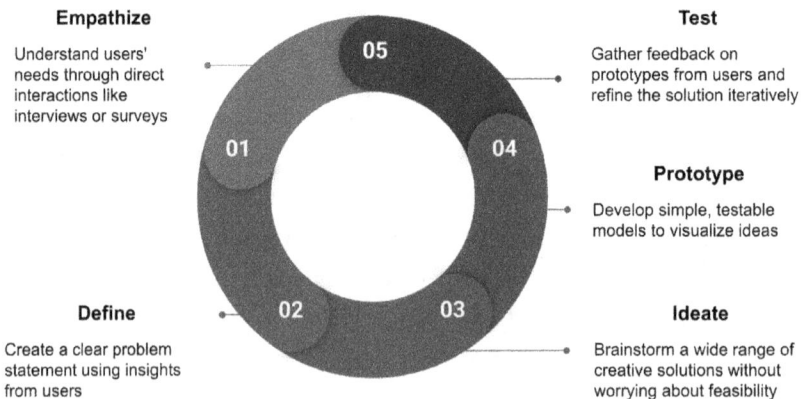

Empathize

Understand users' needs through direct interactions like interviews or surveys

05

01

04

Define

Create a clear problem statement using insights from users

02

03

Test

Gather feedback on prototypes from users and refine the solution iteratively

Prototype

Develop simple, testable models to visualize ideas

Ideate

Brainstorm a wide range of creative solutions without worrying about feasibility

This iterative design process is a flexible and powerful way of developing new ideas to address real-world problems. It can be used in many contexts, including designing and developing new products, developing strategies for new businesses, and innovating solutions for social problems.

Design thinking is well-aligned with a growth mindset, which values learning, flexibility, and the view that abilities can be cultivated through effort. Both leverage the challenges to learn, treat failure as a learning opportunity, and advocate iterative experimentation and refinement. Prototyping and testing invite failure—not as an end but as a means to the end of learning what does not work, which is the growth mindset's way of learning from mistakes. Both focus on the power of feedback and collaboration, which fuels a culture that tends toward openness, responsiveness, and learning.

Mixing design thinking with a growth mindset allows for better innovation and problem-solving results, with an increased focus on the needs of users. With these two combined, teams can iterate and reiterate a solution based on user feedback, creating a more effective and user-centered solution. The power of the growth mindset also cultivates the spirit of experimentation without the fear of failure. Therefore, experimentation encourages creativity and innovation, allowing us to achieve novel and breakthrough innovations.

The iterative nature of the design thinking process allows for solutions to continuously improve and evolve because the needs and wants of the users are subject to change. Thus, design thinking and a growth mindset work together to enable resilience, collaboration, and success in product development, organizational growth, and evolution.

An excellent example of design thinking in practice is the development of the first iPhone. In the early 2000s, mobile phones were large and cumbersome and were not intuitive to use. Apple designers empathized with these end-user frustrations and defined the user-experience problem. They then ideated a disruptive solution: a touchscreen phone. They prototyped and tested some concepts, refining the phone design until they eventually released it as the iPhone in June 2007. The product was simple, intuitive, and easy to use, revolutionizing the mobile phone industry. This was a design-thinking innovation at work.

Apple put the user at the center of their design process. Thanks to this, the company could not only solve practical issues but also create a new paradigm for mobile technology, which was shortly adopted by all the main players in the market. The iterative process allowed Apple to refine the design based on real feedback, ultimately producing a device that transformed how we communicate, access information, and interact with technology daily. What is also worth noting is that this approach demonstrated the power of cross-disciplinary collaboration: managers, engineers, designers, and marketers all worked together to create a product that was both technically advanced and emotionally appealing to consumers.

Chapter 9

Product-Driven Strategies for Advancing Learning in Projects

Implementing practices from product management to the project management cycle is a great way to build a growth mindset. Demos, pilots, experiments, research, and building MVPs help facilitate learning, flexibility, and innovation—traits at the heart of a growth mindset. With the help of the product management area, teams can experiment, absorb constructive feedback, and make data-driven decisions so that projects will never become stagnant. The result is that teams learn to approach obstacles as opportunities to progress, not failures. Moreover, it creates a collaborative, agile culture that is open to constant change based on real-time data so as to ensure the project provides the highest value possible and stays responsive to evolving requirements. Ultimately, when these practices are implemented, uncertainty can be handled confidently, and teams will achieve better performance.

Experiments in Adaptive Project Management

Product experimentation entails testing new features, designs, or other modifications within a product based on user research. The goal is to determine whether these changes enhance user experience, engagement, and various other metrics. This approach employs a systematic and data-driven mindset, involving the continuous design and execution of experiments to facilitate more informed product decisions.[105]

Experiments are essential for teams of product managers and project managers who want to scientifically evaluate different strategies and approaches. Experiments can provide deeper insights into both solutions and the people who are at the heart of those solutions. Doing so can help managers make choices based on what they know rather than what they assume, increasing their chances of bringing projects to completion.

The experiment begins with a clearly defined objective, or a question and a hypothesis built on that objective, about what you think the outcome will be. The next step is to design the experiment through a control group that follows along throughout the current process and a treatment group that uses the new feature or tool. At that point, the experiment is implemented, and its effects on project outcomes are measured with data collected and analyzed using metrics, drawing conclusions from the analysis. The final step is to iterate and learn by applying what was learned to new projects, implementing changes that work, and seeing what else can be tested for improvement.

Benefits of Experiments

- **Data-driven:** Helps project managers make better decisions using real data, taking the guesswork out of project decisions and reducing risk.
- **Continuous improvement:** Enable iterative improvements in project processes, resulting in better processing of the project over time.
- **Stakeholder-centric approach:** Analyze stakeholders' needs and optimize the project delivery to meet their expectations.
- **Ensuring a realistic outcome:** Verify assumptions about deadlines, budgets, and project execution methods, ensuring the feasibility of the outlined plan.

Major Types of Experiments

Below is a range of experiments that can be used in project and product management.[106] Choosing the right one depends on the specific project objective.

- **A/B testing:** Contrasts two variations of a feature to determine the superior performer.
- **Prototype testing:** Employs prototypes to collect feedback and pinpoint areas for enhancement.
- **Pricing experiments:** Assess different pricing models and choices to discover the optimal strategy.
- **UX testing:** Analyzes the product's interface through user testing.
- **Metric optimization:** Enhances conversion rates, onboarding processes, or messaging.
- **Channel and localization experiments:** Test marketing channels or tailor products for diverse markets.
- **Retention and referral experiments**: Investigates tactics to enhance retention, engagement, and referrals.
- **SEO and social media experiments:** Experiments with SEO (Search Engine Optimization) and social media strategies for heightened visibility and engagement.
- **Usability testing:** Recognizes and resolves usability concerns.

Strategic Implementation Through Pilots

Pilots offer teams a chance to gradually expose improvements to a small subset of users before a full rollout. Whether validating a change to an existing feature or a high-risk innovation, they enable a team to study user interactions in a controlled environment.

Pilots are useful to validate the product's performance, collect feedback, and test assumptions. The learning harvested from pilots about user needs, preferences, and behaviors better equips organizations for subsequent iterations.

Key Aspects of Pilots in Project and Product Management

- **Purpose:** Pilots serve as a test to review how the product is received by users or to see if the product is even feasible. It enables product groups to verify their assumptions about users, conduct usability studies on their products, and verify that customer needs are met.

- **Scope:** The pilots' users are carefully selected so that the development team can observe their actions. Users are granted a specific type of access to the product or feature to allow the team control of the way they use it.

- **User feedback:** Pilot participants—whether early adopters, heavy users of the product, or a focus group—are integral to this feedback process, providing valuable information about user experience, features, and improvements.

- **Duration:** Pilots are designed to generate in a short timeframe the exact amount of data needed to allow a decision to be made. They might run for only a couple of days, weeks, or months before the product is ready to roll out to the larger market, depending on the scale and complexity of the tested product.

- **Constructive cycle:** During and after the pilot, the team receives insights that can be qualitative (eg., recommendations) or quantitative (e.g., regarding performance or usage).

- **Outcome:** After a pilot is concluded, the product team checks whether their objectives have been met. If a pilot is successfully completed, the product will go into a larger release phase. If a pilot fails, changes can then be made to resolve the issue, or a strategic pivot can be considered.

- **Risk management:** Pilots play a significant role in the risk mitigation required for new product launches. They serve as a testing mechanism that can prevent releasing a partially functional product or even avoid a disaster during a full-scale rollout.

Pilots provide a priceless opportunity for companies to road-test ideas with potential customers before making products available for larger segments of customers. By carefully selecting small groups of end users to trial solutions, product teams can gain a profound understanding of the situation without inflating costs or stretching timelines. Whether uncovering bugs, assessing how well functions meet needs, or collecting candid feedback, pilots provide opportunities to substantially refine the value that solutions deliver and identify what risks remain before rolling out at scale.

With iterative changes informed by early adopter feedback and usage data, pilot programs can help ensure that solutions genuinely solve problems as intended and provide experiences worth using continuously once fully implemented across organizations. For some pilots, converging viewpoints reveal where simplification might result in increased adoption without compromising benefits. For other pilots, converging viewpoints lead to a re-examination of assumptions and the question of whether tweaking technical or economic models could result in a broader solution revision. In all cases, pilots accumulate insights that can inform the next iteration of the solution, making it more suitable to meet actual user needs.

Beta Testing: Evaluating Innovations

Testing in product and project management refers to the releasing of a near-ready product to a limited group of external users to test its functionality, usability, and performance in real-life conditions before the official launch.

The beta testing process starts with picking beta testers, who are, in most cases, users of the target market or just volunteers familiar with the domain of the product. Testers receive a beta version of the product, as well as supplementary instructions and guidance on how to use the product in the most effective way, and are requested to report any bugs, glitches, or usability issues they encounter.

Beta users will most likely use the tested product or feature a multitude of environments—a diverse set of testing conditions that the internal team might not have foreseen and which will reveal technical bugs, performance issues, and design flaws that might have otherwise gone unnoticed. The product team establishes feedback channels that allow users to quickly and easily voice their thoughts about the product while they are using it: surveys, forums, or other direct communication channels to collect detailed, granular feedback about how the product is behaving and what aspects of it might still need improving.

This technique also involves watching key performance metrics such as load times, crashes, and error reports to make sure the product is stable. The development team analyzes this feedback, prioritizes bug fixes, and makes final adjustments based on user and data-driven feedback.

Beta Testing vs Pilots

Although beta testing and pilots are both essential in project and product management, they differ in many respects. Beta testing is more appropriate during development. It happens later in the process and is when the technical bugs are usually removed and usability is improved. This testing is typically run with a larger group of external users, and the product is nearly fully released.

Pilots, on the other hand, happen earlier in the process. Their main objective is to validate the product, the business, and the consumers. Pilots are not about the technical quality of the product but rather about the 'feasibility' of the project. They involve a smaller and more controlled group and are used to assess more abstract factors, such as product-market fit or the business model. Beta testing is for technical quality, while pilots are to evaluate whether something can be brought to scale and thus commercialized.

The key differences between these two techniques are presented in the following table.

Table 19. Beta Testing vs Pilots

Aspect	Beta Testing	Pilot
Main Goal	Bugs or issue identification, reviewing technical readiness.	Product business validation, feasibility check, collecting insights.
Scope	Primarily narrow, focused on the technical aspect.	Primarily broad, focused on business and user aspects.
Audience	A limited group of external end users.	Small group of users or customers, selected based on defined criteria.
Feedback	Technical; focused on usability.	Business insights; focused on user experience.
Timing	At the end of the product development lifecycle.	Can be conducted at every stage of the product development lifecycle.
Following Actions	Bugs fixes, issues resolving.	Checking product against requirements; collecting insights.
Growth Mindset	Continuous improvement.	Fostering feedback culture.

Minimum Viable Products (MVPs) as Growth Tools

The Minimum Viable Product (MVP) is a useful technique for project management with a growth mindset. This approach gained widespread traction in modern project management, leveraging Lean and Agile methodologies. One book that has helped popularize the MVP concept was Eric Ries's *The Lean Startup*. For him, the MVP is the early version of the product that enables a full turn of the Build-Measure-Learn loop with a minimum amount of effort and the least amount of development time.[107]

At its core, MVP refers to the notion that iterative testing and user feedback should guide development more than extensive pre-launch planning. By launching the simplest form of their concept for examination by potential customers, product and project teams can learn whether their underlying assumptions actually resonate with the public, then either pivot or expand upon validated aspects based on these real-world responses.

This minimalist approach mirrors Agile software development and Lean manufacturing philosophy by prioritizing iteration before detailed pre-planning. Customer feedback is at the center of product refinement and the assessment of viability. The MVP model has, therefore, become linked with modern product design, preferring the flexible learning model over rigid blueprints with an agile touch; as such, the MVP model has become the darling of the startup community.

MVP helps to rapidly validate ideas and establish a culture of experimentation. With a focus on core features that provide immediate value, MVPs enable teams to quickly test assumptions through interactions with real users. Seeing as what is learned from this interaction arrives back to the product design and development teams, MVPs thus facilitate a customer-oriented approach to product features that balance user requirements and market needs with Agile project methodologies that promote adaptability and rapid response to change.

By reducing inefficient activities and the wasting of valuable time, MVPs allow teams to verify assumptions with actual customers, saving precious resources and safeguarding them from investing heavily in features that may never achieve proof of concept. The MVP process also enables teams to establish a feedback loop that ensures ongoing learning and cooperation.

Once the MVP is validated and scaled, teams can confidently expand with the use of larger resources to reach a wider market. In short, the MVP process bolsters adaptability through its ability to create safe containers for learning and accountability that kickstart fast feedback loops, stop wasting valuable resources on the wrong things, and eliminate the risks associated with forging ahead without constantly checking in. Below is a table summarizing the key benefits of using a Minimum Viable Product.

Table 20. Key Benefits of MVP

Benefit	Description
Rapid Validation	Allows teams to test product ideas quickly with real users, validating assumptions early in the process.
Customer-Centric Development	Focuses on features that address user needs, ensuring products are aligned with customer expectations.
Reduced Risk	Minimizes risks associated with product launches by allowing for adjustments based on user input before a full-scale release.
Resource Optimization	Prevents over-investment in untested features, optimizing time and budget allocation by concentrating on the essentials.
Agile Adaptability	Encourages flexibility and responsiveness to change, facilitating Agile practices within the development process.
Feedback Loop	Establishes a continuous feedback mechanism, enabling teams to learn from users and iterate effectively.
Confident Scaling	Enables teams to confidently scale the product once validated, ensuring it meets market demands effectively.
Growth Mindset	Promotes a culture of learning and experimentation, allowing teams to explore new ideas without fear of failure.

Integrating CX and UX Research into Project Frameworks

Customer Experience (CX) and User Experience (UX) research play a critical role in product and project management. CX research and UX research both aim to understand how users engage with a product or service and ensure that it is user-friendly and meets their needs while providing a positive experience at every step of the customer journey. Both CX and UX research deliver critical insights that drive business decisions.

The difference between the two types of research lies in their extent: UX research focuses on the product itself and how it is used ("Is it intuitive?"; "Can users complete critical tasks?"), while CX focuses on the entire customer journey ("What's the overall perception of the product or feature?"). Both types of research are important for product management: UX research ensures that the product is usable and meets user expectations, while CX research ensures that the product is aligned with broader customer expectations so that the customer continues to use the product and remains loyal. Working together, these two types of research allow product teams to develop solutions that satisfy user needs and provide value.

Project managers can also benefit from UX and CX research to ensure their project team creates outcomes that meet customer needs and expectations. UX research ensures that the project team understands user needs during development, while CX research ensures that the project team understands customer expectations every step of the way. Together, these two research strategies help mitigate risk, keep the project on track, and ensure that the final outcome meets customer expectations.

Building on these two research strategies also naturally leads to continuous improvement and the nurturing of a growth mindset. Continuous improvement also depends on feedback. Each piece of feedback from users is an opportunity for improvement, enhancing the product or feature and responding to customer needs.

The following table sums up the various differences and similarities between CX and UX research, displaying the role each plays in product and project management while highlighting their contribution to improving productivity and achieving a growth mindset.

Table 21. UX Research vs CX Research

Aspect	User Experience (UX) Research	Customer Experience (CX) Research
Focus Area	Centers on product usability and functionality.	Centers on the customer journey and the overall interactions with the brand.
Key Persona	Primarily user.	Primarily customer.
Goal	Design solutions that are centered on user needs.	Create a cohesive and positive experience across all interactions with the organization.
Scope	Addresses specific interactions within the product.	Takes a broader view of the product, including marketing, sales, and support.
Importance in Project / Product Management	Informs decision-making by focusing on user requirements. Ensures the product is intuitive and meets user needs.	Ensures the project meets customer expectations across all touchpoints. Aligns the product with overall customer expectations, enhancing loyalty.
Growth Mindset	Promotes learning and continuous improvement based on user insights.	Views feedback as valuable and sees challenges as opportunities for growth.

Methods and Techniques Used in UX and CX research

In general, research methods are divided into quantitative and qualitative ones. Quantitative research is deductible, objective, and has a wider scope, while qualitative research is inductive, subjective, and narrower in scope.[108]

Qualitative research is interested in learning about things that cannot be quantified (e.g., experiences, groups), and relies on sources of data that are nonnumerical (e.g., video footage, interviews). Quantitative research, on the other hand, aims to provide exact, measurable, causal explanations, usually in a controlled environment in which the data collection method is objective and involving a type of experimentation that answers specific questions in a precise mathematical way.

- **CX Qualitative Research** digs into why customers think, feel, and engage with a brand or service in specific ways. Key methods include customer journey mapping, interviews, focus groups, and ethnographic studies. Example outputs from this approach include customer insights and context obtained about a customer's experience and expectations.
- **CX Quantitative Research** is the approach to measuring customer satisfaction and loyalty using numbers. The process includes the collection and analysis of both internal and external data from various sources, such as surveys (e.g., Net Promoter Score or Customer Sentiment Score) or analytics tools. The goal is to quantify the experience of the customer, identify trends, and provide an objective metric that can inform business decisions.
- **UX Qualitative Research** tries to uncover aspects such as user experience, motivations, and behaviors with thick descriptions. Common qualitative approaches include user interviews, contextual inquiries, usability testing, focus groups, and diary studies. The priority is on 'why' and attempting to add a layer of richness to the user's interactions.
- **UX Quantitative Research** collects and analyzes numeric data such as user behaviors and preferences. The methods to attain this data include surveys, A/B testing, analytics, and heat maps. Based on these data points, it is possible to quantify experiences and uncover trends to provide business cases for design decisions with measurable outcomes.

Demos as a Communication and Validation Tool

A demo is a presentation by which the project team demonstrate the features, benefits, and applications of a product or software. The objective is to let people know how to use your product and why it is a viable alternative for them. Demo can be used to present the final product to customers or review the progress of product development internally within the organization.[109]

Internal demos are especially important for early support-building for a new product. They help show recent progress and get feedback from key internal stakeholders or customers as the product is being developed and tested. In Agile methodologies, demos are extensively used to showcase incremental deliverables regularly to critical stakeholders for validation against requirements and business needs.

Delivering demos supports a growth mindset culture by enabling learning and feedback, as well as continuous improvement.

Advantages of Demos

- **Stakeholder alignment and live feedback:** Providing demos ensures everyone is on the same page about the final deliverable or the current state of the product/feature development. Cross-team checks are performed to obtain feedback across different areas of the company.
- **Risk mitigation:** Reviewing a demo decreases the likelihood that you will release a product or feature with a bug. The demo can itself become an additional QA (Quality Assurance)—the more times you show a deliverable to cross-functional stakeholders, the more chances there are to identify possible issues or negative impacts on user experience.
- **Communication:** Demos enhance communication. They are an effective way to communicate changes, developments, and updates with your fellow silos and departments so that they have the opportunity to ask questions, provide feedback, and

assess how these changes impact the overall customer-facing changes and delivered value.

- **Team motivation:** Product demos create a sense of responsibility and commitment. Since their work will be presented to stakeholders, the team must now focus on the quality of the product and its timely delivery. Demos include a distinct deadline and encourage collaboration and task prioritization to meet the established standards. The expectation of showing a working, functional product creates a sense of pride in their work, which in turn motivates team members to work even harder.

Community Engagement

For product and project managers, building a customer community is a goldmine of feedback: it provides unfettered access to customers' unfiltered responses. Customer communities are groups of customers organized by companies to keep in contact and engage with users of their products. These groups are meticulously designed to gather feedback on a structured medium, which in turn helps companies build a better understanding of their customer's needs and wants. Such insights can help product teams improve the usability of their products, prioritize features, and identify opportunities for improvement. Customer communities are also the best place to watch for risks and provide support so that product teams can identify and fix issues that community members raise regarding product functions.

If product managers follow these discussion threads in the community, they can be kept abreast of market trends and understand the performance of their products in real time. To put it simply, customer communities empower product managers and project teams to make a more informed decision based on detailed analysis of the unfiltered views emerging from the discussion in the community, leading to improved products that better meet customers' needs.

Communities may also include product training materials and courses, bringing additional value for customers and promoting the culture of learning.

Below are some of the benefits the community may create for the organization:

- **Feedback and insights:** Customer communities are a natural place for gathering feedback on existing products and features and gauging interest in new products and features. Members can ask questions, share their experiences, learn, provide feedback and suggestions, and highlight pain points to help product managers better understand user needs.

- **Customer-centric development:** Iterating with users empowers product managers to be more customer-centric in their product development process. By understanding user tastes and habits, teams can prioritize features that will have the most impact.

- **Idea validation:** With community members, concepts can be validated prior to launching a feature or product, thus reducing the risk of investing in features that might not be meaningful.

- **Building advocacy:** Bringing customers into the brand's community makes them feel more included and leads to greater loyalty in the long run. Users who are more invested in a product tend to act as evangelists for the brand or product.

- **Co-creating:** Communities can be a form of co-creation whereby users collaborate with product teams to design and refine features—improving the quality of products and the relationship with users.

- **Trend identification:** If the product manager reads conversations in customer communities, they will notice up-and-coming trends, changing user needs, and opportunities for new markets. The product manager can then act on this information before a competitor does.

- **Support and education:** Users often come together in communities to help each other, generating valuable support, education, and documentation ideas for product managers.
- **Networks and partnerships:** Communities can link users with common interests or requirements, opening up potential networking and partnership possibilities that can influence the product ecosystem.

Notes

Introduction

[1] McKinsey & Company, *The State of Organizations 2023,* accessed November 26, 2024, https://www.mckinsey.com/~/media/mckinsey/business%20functions/people%20and%20organizational%20performance/our%20insights/the%20state%20of%20organizations%202023/the-state-of-organizations-2023.pdf

[2] U.S. Bureau of Labor Statistics, "Occupation Outlook Handbook: Project Management Specialists," accessed November 26, 2024, https://www.bls.gov/ooh/business-and-financial/project-management-specialists.htm

[3] Project Management Institute, *Talent Gap: Ten-Year Employment Trends, Costs, and Global Implications* (2021), accessed December 10, 2024, https://www.pmi.org/-/media/pmi/documents/public/pdf/learning/career-central/talent-gap-report-2021-finalfinal.pdf

[4] Project Management Institute, *Pulse of the Profession® 2024: The Future of Project Work: Moving Past Office-Centric Models*, 15th ed., accessed November 26, 2024, https://www.pmi.org/-/media/pmi/documents/public/pdf/learning/thought-leadership/pmi-pulse-of-the-profession-2024-report.pdf

[5] Ibid, 13.

Chapter 1

[6] Dweck, Carol S., *Mindset: The New Psychology of Success* (New York: Ballantine Books, 2016).

[7] Murphy, Mary C., *Cultures of Growth: How the New Science of Mindset Can Transform Individuals, Teams, and Organizations* (New York: Simon & Schuster, 2024).

[8] Harvard Business Review, "How Companies Can Profit from a 'Growth Mindset,'" *Harvard Business Review*, November 2014, accessed October 19, 2024, https://hbr.org/2014/11/how-companies-can-profit-from-a-growth-mindset

[9] Wolfe, Ira, "Five Business Benefits of Growth Mindset: How to Thrive in Today's Competitive Environment," *Forbes*, May 5, 2022, 2024, https://www.forbes.com/councils/forbescoachescouncil/2022/05/05/five-business-benefits-of-growth-mindset-how-to-thrive-in-todays-competitive-environment/

[10] Milkman, Katherine, "Giving Feedback That Works: Plant the Seeds of Confidence," *Knowledge at Wharton*, January 24, 2022, https://knowledge.wharton.upenn.edu/article/giving-feedback-works-plant-seeds-confidence/

Chapter 2

[11] Fast Company, "Coke's Freestyle Machine is an IoT Evangelist," March 22, 2016, https://www.fastcompany.com/3058161/cokes-freestyle-machine-is-an-iot-evangelist

[12] Doering, Christopher, "How Coca-Cola Turns to Its Freestyle Machine to Create Shelf-Ready Flavors", FoodDive, December 15, 2022, https://www.fooddive.com/news/how-coca-cola-turns-to-its-freestyle-machine-to-create-shelf-ready-flavors/636660/

[13] Ibid.

[14] Project Management Institute, *A Guide to the Project Management Body of Knowledge (PMBOK® Guide)*, 7th ed. (Newtown Square, PA: Project Management Institute, 2021), 55.

[15] Oxlade, Chris, *The Light Bulb (Tales of Invention)*, (Chicago, IL: Heinemann Library/Capstone, 2012).

[16] Olson, DaiWai M., "In Research, Failure Is Not an Option—It Is an Expectation," *Journal of Neuroscience Nursing* 55, no. 6 (December 2023): 187, https://doi.org/10.1097/JNN.0000000000000734

[17] Satell, Greg, "It's Time to Bury the Idea of the Lone Genius Innovator," *Harvard Business Review*, April 6, 2016, https://hbr.org/2016/04/its-time-to-bury-the-idea-of-the-lone-genius-innovator

[18] Horth, David Magellan and Vehar, Jonathan, *Becoming a Leader Who Fosters Innovation* (Center for Creative Leadership, 2014), 17. Accessed October 24, 2024, https://www.ccl.org/wp-content/uploads/2014/03/BecomingLeaderFostersInnovation.pdf

[19] Slack Technologies, *"Slack is Where More Work Happens Every Day, All Over the World"*, January 29, 2019, https://slack.com/blog/news/slack-has-10-million-daily-active-users

[20] Slack Technologies, "Slack Announces Strong Fourth Quarter and Fiscal Year 2021 Results," March 4, 2021, https://slack.com/blog/news/slack-announces-strong-fourth-quarter-and-fiscal-year-2021-results

[21] Netflix, "About Netflix," accessed October 7, 2024, https://about.netflix.com/en

[22] LEGO Group, "The LEGO Group Delivers Double-Digit Top- and Bottom-Line Growth in H1 2024," August 28, 2024, https://www.lego.com/en-sg/aboutus/news/2024/august/The-LEGO-Group-delivers-double-digit-growth-in-H1-2024

[23] Project Management Institute, *A Guide to the Project Management Body of Knowledge (PMBOK® Guide)*, 7th ed. (Newtown Square, PA: Project Management Institute, 2021), 55.

Chapter 3

[24] Murphy, *Cultures of Growth*, 7.

[25] Canning, Elizabeth A., Murphy, Mary C., Emerson, Katherine T. U., Chatman, Jennifer A., Dweck, Carol S., and Kray, Laura J., "Cultures of Genius at Work: Organizational Mindsets Predict Cultural Norms, Trust, and Commitment," *Personality and Social Psychology Bulletin* 46, no. 4 (2020): 626–642, September 10, 2019, https://doi.org/10.1177/0146167219872473

[26] Dweck, Carol, and Kathleen Hogan, "How Microsoft Uses a Growth Mindset to Develop Leaders," *Harvard Business Review*, October 2016, https://hbr.org/2016/10/how-microsoft-uses-a-growth-mindset-to-develop-leaders

[27] Microsoft, "Microsoft Mission," accessed December 1, 2024, https://www.microsoft.com/en-us/about

[28] Ibid.

[29] Sachs, Aaron, and Kundu, Anupam, *The Unfinished Business of Organizational Transformation* (Thoughtworks, 2014), November 13, 2015, https://www.thoughtworks.com/insights/blog/unfinished-business-organizational-transformation

[30] Crabtree, Steve, "Worldwide, 13% of Employees Are Engaged at Work," *Gallup*, October 8, 2013, https://news.gallup.com/poll/165269/worldwide-employees-engaged-work.aspx

[31] Ignatius, Adi, "How Indra Nooyi Turned Design Thinking into Strategy: An Interview with PepsiCo's CEO," *Harvard Business Review*, September 2015, https://hbr.org/2015/09/how-indra-nooyi-turned-design-thinking-into-strategy

[32] Budak, Alex, *Becoming a Changemaker: An Actionable, Inclusive Guide to Leading Positive Change at Any Level* (New York: Balance, 2022).

[33] Lima, Antonieta and Riesenberger, Carlos, *Six Sigma Philosophy* (n.p., 2020), 4–6.

[34] Hastwell, Clair, "What Are Employee Resource Groups (ERGs)?," *Great Place to Work*, January 7, 2023, https://www.greatplacetowork.com/resources/blog/what-are-employee-resource-groups-ergs

[35] Gotian, Ruth, and Andy Lopata, *Financial Times Guide to Mentoring* (Harlow, England: Pearson, 2024), 6–7.

[36] Zucker, Rebecca, "The 4 Types of Mentors and the Best Kind of Mentoring Relationships," *Forbes*, July 8, 2004,

https://www.forbes.com/sites/rebeccazucker/2024/07/08/types-of-mentors-and-ideal-mentoring-relationships/

[37] Gotian, Ruth, and Andy Lopata, *Financial Times Guide to Mentoring*, 6–7.

[38] Cartolari, Rose, "Want Your Team to Adopt a Growth Mindset? Here's How to Start with Yourself," *Forbes*, December 5, 2018, https://www.forbes.com/councils/forbescoachescouncil/2018/12/05/want-your-team-to-adopt-a-growth-mindset-heres-how-to-start-with-yourself

[39] Minkin, Rachel, "Diversity, Equity and Inclusion in the Workplace," *Pew Research Center*, May 17, 2023, https://www.pewresearch.org/social-trends/2023/05/17/diversity-equity-and-inclusion-in-the-workplace

[40] Chadha, Nisha, "Teaching 2.0: What's Growth Mindset and Psychological Safety Got to Do with It?," *Journal of Academic Ophthalmology* 12, no. 1 (2020): e20, https://doi.org/10.1055/s-0040-1703017

Chapter 4

[41] Phillipy, Mark A., "Delivering Business Value: The Most Important Aspect of Project Management," paper presented at PMI® Global Congress 2014—North America, Phoenix, AZ (Newtown Square, PA: Project Management Institute, 2014), https://www.pmi.org/learning/library/delivering-business-value-9378

[42] Project Management Institute, *Value Delivery System Explainer*, accessed October 28, 2024, https://www.pmi.org/-/media/pmi/documents/public/pdf/pmbok-standards/value-delivery-system-explainer.pdf

[43] Project Management Institute, *Pulse of the Profession® 2020: Ahead of the Curve: Forging a Future-focused Culture*, 3, accessed October 30, 2024, https://www.pmi.org/-/media/pmi/documents/public/pdf/learning/thought-leadership/pulse/pmi-pulse-2020-final.pdf

[44] Monday.com, "Top Project Management Statistics," February 1, 2022, https://monday.com/blog/project-management/project-management-statistics

[45] Project Management Institute, *A Guide to the Project Management Body of Knowledge (PMBOK® Guide)*, 7th ed. (Newtown Square, PA: Project Management Institute, 2021), 117.

[46] Ibid.

[47] Jordan, Andy, *Risk Management for Project Driven Organizations: A Strategic Guide to Portfolio, Program and PMO Success* (New York: Routledge, 2014), 29–30.

[48] Hillson, David, *Managing Risk in Projects* (London: Routledge, 2009), 7.

[49] De Meyer, Arnoud, Loch, Christoph H. and Pich, Michael T. "Managing Project Uncertainty: From Variation to Chaos," *MIT Sloan Management Review*, January 15, 2002, https://sloanreview.mit.edu/article/managing-project-uncertainty-from-variation-to-chaos

[50] Workfront, *Measuring and Analyzing Work* (2017), accessed October 2, 2024, https://business.adobe.com/resources/articles/measure-and-analyzing-work/thank-you.html

[51] Kaplan, Robert S. and Norton, David P., "The Office of Strategy Management," *Harvard Business Review*, October 2005, https://hbr.org/2005/10/the-office-of-strategy-management

[52] Sull, Donald, Homkes, Rebecca and Sull, Charles, "Why Strategy Execution Unravels—And What to Do About It," *Harvard Business Review*, March 2015, https://hbr.org/2015/03/why-strategy-execution-unravelsand-what-to-do-about-it

[53] Locke, Edwin A. and Latham, Gary P., "Building a Practically Useful Theory of Goal Setting and Task Motivation: A 35-year Odyssey." *American Psychologist* 57, no. 9 (2002): 705–717, https://psycnet.apa.org/doi/10.1037/0003-066X.57.9.705

Chapter 5

[54] Project Management Institute, *Pulse of the Profession® 2024: The Future of Project Work: Moving Past Office-Centric Models*, 15th ed., accessed November 26, 2024, https://www.pmi.org/-/media/pmi/documents/public/pdf/learning/thought-leadership/pmi-pulse-of-the-profession-2024-report.pdf

[55] Radigan, Dan, "Agile vs. Waterfall Project Management," *Atlassian*, accessed October 16, 2024, https://www.atlassian.com/agile/project-management/project-management-intro

[56] Digital.ai, *The 17th State of Agile Report* (2024), accessed December 12, 2024, https://digital.ai/press-releases/17th-state-of-agile-report-71-use-agile-in-their-sdlc-small-organizations-report-strong-business-benefits-medium-and-larger-sized-companies-continue-to-experience-barriers-in-successfully-scaling-a

[57] *Manifesto for Agile Software Development*, accessed October 1, 2024, https://agilemanifesto.org

[58] PM Solutions, *The State of the Project Management Office (PMO) 2022*, 4, accessed November 28, 2024, https://www.pmsolutions.com/articles/Project_Management_2022_Research_Report.pdf

[59] Whitaker, Sean, "The Benefits of Tailoring: Making a Project Management Methodology Fit," *PMI White Paper*, September 2014, https://www.pmi.org/learning/library/tailoring-benefits-project-management-methodology-11133

[60] Ibid.

[61] KPMG, *2022 KPMGxPMI Project Management Survey Report* (KPMG, 2023), 12, accessed November 21, 2024, https://assets.kpmg.com/content/dam/kpmg/cy/pdf/2023/kpmg_pmi_project_management_survey_2022.pdf

Chapter 6

[62] Laker, Benjamin, "How to Manage Stakeholder Expectations for Better Outcomes," *Forbes*, June 3, 2024, https://www.forbes.com/sites/benjaminlaker/2024/06/03/how-to-manage-stakeholder-expectations-for-better-outcomes/

[63] Project Management Institute, *A Guide to the Project Management Body of Knowledge (PMBOK® Guide)*, 7th ed. (Newtown Square, PA: Project Management Institute, 2021), 12–13.

[64] Baker, Colin, "7 Types of Feedback for the Workplace (and One to Avoid)," *Leaders Media*, August 31, 2022, https://leaders.com/articles/business/types-of-feedback

[65] Society for Human Resource Management (SHRM), "The Importance of Civility in the U.S." accessed November 23, 2024, https://www.shrm.org/topics-tools/topics/civility

[66] Harter, Jim, "In New Workplace, U.S. Employee Engagement Stagnates," *Gallup*, January 23, 2024, https://www.gallup.com/workplace/608675/new-workplace-employee-engagement-stagnates.aspx

[67] The Conference Board, "DNA of Engagement: How Organizations Create and Sustain Highly Engaged Teams?," February 28, 2019, https://www.conference-board.org/publications/how-organizations-create-sustain-highly-engaged-teams-report

[68] Saundry, Richard, and Unwin, Peter, "Estimating the Costs of Workplace Conflict," *Acas*, May 11, 2021, https://www.acas.org.uk/research-and-commentary/estimating-the-costs-of-workplace-conflict/report

[69] CPP, *Workplace Conflict and How Businesses Can Harness It to Thrive*, CPP Global Human Capital Report, July 2008, https://shop.themyersbriggs.com/Pdfs/CPP_Global_Human_Capital_Report_Workplace_Conflict.pdf

[70] Bennett, Michelle, "Workplace Conflict Statistics: How We Approach Conflict at Work," *Niagara Institute*, August 11, 2022, https://www.niagarainstitute.com/blog/workplace-conflict-statistics

[71] Bravely, "Understanding the Conversation Gap: Why Employees Aren't Talking, and What We Wan Do About It," accessed November 21, 2024, https://learn.workbravely.com/hubfs/Understanding-the-Conversation-Gap.pdf

[72] Ibid.

[73] Knight, Rebecca, "How to Handle Difficult Conversations at Work," *Harvard Business Review*, January 9, 2015, https://hbr.org/2015/01/how-to-handle-difficult-conversations-at-work

Chapter 7

[74] EY, "EY Survey Reveals Huge Uptick in GenAI Adoption at Work, Correlates with 'Talent Health' and Competitive Gains," October 15, 2024, https://www.ey.com/en_gl/newsroom/2024/10/ey-survey-reveals-huge-uptick-in-genai-adoption-at-work-correlates-with-talent-health-and-competitive-gains

[75] PwC, "PwC's 27th CEO Survey: Thriving in an Age of Continuous Reinvention," January 15, 2024, https://ceosurvey.pwc

[76] Eloundou, Tyna, Manning, Sam, Mishkin, Pamela and Rock, Daniel, "GPTs are GPTs: An Early Look at the Labor Market Impact Potential of Large Language Models," *arXiv*, August 23, 2023, https://arxiv.org/abs/2303.10130

[77] PwC, "Sizing the Prize. PwC's Global Artificial Intelligence Study: Exploiting the AI revolution," 2017, accessed December 17, 2024, https://www.pwc.com/gx/en/issues/artificial-intelligence/publications/artificial-intelligence-study.html.

[78] Pomeroy, Robin and Myers, Joe, "AI—Artificial Intelligence—at Davos 2024: What to Know," *World Economic Forum*, January 14,

2024, https://www.weforum.org/stories/2024/01/artificial-intelligence-ai-innovation-technology-davos-2024

[79] Project Management Institute, *Shaping the Future of Project Management With AI*, October 2023, https://www.pmi.org/learning/thought-leadership/ai-impact/shaping-the-future-of-project-management-with-ai

[80] KPMG, "Artificial intelligence (AI) in the World of Project Management," September 2021, https://kpmg.com/ae/en/home/insights/2021/09/the-evolving-role-of-the-project-manager.html

[81] Loutfi-Hipchen, Elizabeth, "Revolutionizing Learning: The Power of AI and VR in Employee Development," *Chief Learning Officer*, February 29, 2024, https://www.chieflearningofficer.com/2024/02/29/revolutionizing-learning-the-power-of-ai-and-vr-in-employee-development

[82] IBM, *The Cognitive Enterprise: Reinventing your Company with AI. Seven Keys to Success*, February 2019, https://www.ibm.com/downloads/cas/GVENYVP5

[83] Nielsen, Jakob, "AI Improves Employee Productivity by 66%," NN Group, July 16, 2023, https://www.nngroup.com/articles/ai-tools-productivity-gains

[84] BrandPartner, "Unleashing the Power of AI in the Workplace: How It Might Impact You?", June 19, 2023, https://brandpartner.id/unleashing-the-power-of-ai-in-the-workplace-how-it-might-impact-you

[85] Edlich, Alexander, Ip, Fanny and Whiteman, Rob, "How Bots, Algorithms, and Artificial Intelligence Are Reshaping the Future of Corporate Support Functions," *McKinsey Digital*, November 15, 2018, https://www.mckinsey.com/capabilities/mckinsey-digital/our-insights/how-bots-algorithms-and-artificial-intelligence-are-reshaping-the-future-of-corporate-support-functions

[86] Robinson, Bryan, "Employees View AI as an Asset, Not a Threat, According to New Study," Forbes, July 7, 2023, https://www.forbes.com/sites/bryanrobinson/2023/07/07/employees-view-ai-as-an-asset-not-a-threat-according-to-new-study

[87] Michelen, Abe, "Between 400 and 800 Million Jobs Lost to Automation by 2030: McKinsey," *Towards Data Science*, November 30, 2017, https://insights.globalspec.com/article/7278/between-400-and-800-million-jobs-lost-to-automation-by-2030-mckinsey

[88] Farrel, Robert, "The Impact of AI on Job Roles, Workforce, and Employment: What You Need to Know," *Innopharma Education*, September 29, 2023, https://www.innopharmaeducation.com/blog/the-impact-of-ai-on-job-roles-workforce-and-employment-what-you-need-to-know

[89] Hendershot, Steve, Bobburi, Naveen Goud and Li, Lea. "Why AI Ethics Matters in Project Management," *Projectified® Podcast*, Project Management Institute, December 7, 2024, https://www.pmi.org/projectified-podcast/podcasts/why-ai-ethics-matters-in-project-management

Chapter 8

[90] Digital.ai, *The 17th State of Agile Report* (2024), 11.

[91] Pries, Kim H. and Quigley, Jon M., *Agile Project Management with Scrum* (Boca Raton: CRC Press, Taylor and Francis Group, 2011), 2.

[92] Anderson, David J. and Reinertsen, Donald G., *Kanban: Successful Evolutionary Change for Your Technology Business* (Sequim, WA: Blue Horse Press, 2010), 15.

[93] U.S. Environmental Protection Agency, "Lean Thinking and Methods—5S," accessed October 12, 2024, https://www.epa.gov/sustainability/lean-thinking-and-methods-5s

[94] Kaizen Institute, "KAIZEN™ Approach to Operations: Principles and Model," accessed October 7, 2024, https://kaizen.com/insights/kaizen-operations-principles

[95]Alukal, George and Manos, Anthony, *Lean Kaizen: A Simplified Approach to Process Improvements* (Milwaukee: ASQ Quality Press, 2006), 10.

[96] Kato, Isao and Smalley, Art, *Toyota Kaizen Methods: Six Steps to Improvement* (Boca Raton: CRC Press, Taylor and Francis Group, 2010), 4–5.

[97] LeanSuite, "The Five Elements of Kaizen," May 31, 2022, https://theleansuite.com/the-five-elements-of-kaizen

[98] Janjić, Vesna, Radoslav Bogicevic, Jasmina and Krstić, Bojan, "Kaizen as a Global Business Philosophy for Continuous Improvement of Business Performance," *Ekonomika* 65, no. 2 (April–June 2019): 17, https://doi.org/10.5937/ekonomika1902013J

[99] Myszewski, Jan M., "On Improvement Story by 5 Whys," *The TQM Journal* 25, no. 4 (June 2013): 371, https://doi.org/10.1108/17542731311314863

[100] Kaizen Institute, "Understanding Continuous Improvement: A Guide for Operational Excellence," accessed October 15, 2024, https://kaizen.com/insights/continuous-improvement-operational-excellence

[101] Baird, Craig, *The Six Sigma Manual for Small and Medium Businesses: What You Need to Know Explained Simply* (Ocala: Atlantic Publishing Group Inc., 2009), 19.

[102] Pyzdek, Thomas and Keller, Paul, *The Six Sigma Handbook*, 3rd ed. (New York: McGraw-Hill Professional, 2009), 149.

[103] The Council for Six Sigma Certification, *Six Sigma: A Complete Step-by-Step Guide: A Complete Training & Reference Guide for White Belts, Yellow Belts, Green Belts, and Black Belts* (Buffalo: Harmony Living, 2018), 21, https://www.sixsigmacouncil.org/wp-content/uploads/2018/08/Six-Sigma-A-Complete-Step-by-Step-Guide.pdf

[104] Lima and Riesenberger, *Six Sigma Philosophy*, 3–4.

Chapter 9

[105] Singh, Bhavya, "The Art of Experimentation for Product Managers," *Medium*, August 8, 2023, https://million-rare.medium.com/the-art-of-experimentation-for-product-managers-3c07de94fc7

[106] Ibid.

[107] Ries, Eric, *The Lean Startup: How Today's Entrepreneurs Use Continuous Innovation to Create Radically Successful Businesses* (New York: Crown Publishing Group, 2011), 77.

[108] Oflazoglu, Sonyel (ed.), *Qualitative Versus Quantitative Research* (London, UK: IntechOpen, 2017), 5.

[109] Brennan, Kevin, *Mastering Product Management: A Step-by-step Guide* (n.p., 2019), 33.

List of Tables

List of Figures

Index

Acknowledgments

AI-Assisted Content Acknowledgment

The content in this book was created by Magda Jaworowicz and Peter Jaworowicz. The authors used AI-assisted methods in its creation. Leading AI tools were utilized for research, insight generation, and refinement of the content. AI also contributed to the brainstorming of topics and relevant enhancements, including text editing and error checking. The role of AI remained assistive, and the book's creators composed all the final text, narratives, and perspectives presented herein.

This statement affirms transparency in the creative process and upholds the integrity of the work, aligning with the established practices for using AI responsibly in publishing.

Acknowledgment of Fictional Characters

The examples and characters presented in this book, particularly in Chapters 8 and 9, are entirely fictional and were created by the authors to illustrate the concepts and processes discussed.

Using fictional characters allows for a nuanced exploration of ideas while ensuring confidentiality and clarity in demonstrating key principles. Any resemblance to actual persons, living or deceased, or real-world situations is purely coincidental.

About the Authors

Magda Jaworowicz, PhD, MBA, PMP

Magda (*Magdalena*) Jaworowicz has over 15 years of experience managing marketing, learning, and development programs for global technology companies. Currently, she serves as the Partner Marketing Learning and Enablement Lead at Cisco. Magda holds a PhD in Communication, an MBA in Management, an MA in Education, and two project management certifications: PMP and PRINCE2.

In addition to her expertise in project management, Magda is a certified mindset coach specializing in mindful leadership. She developed the "Grow Your Mindset" program to help corporate professionals foster personal and career growth. This program integrates her extensive experience in global tech, coaching, academia, and nonprofit leadership, with a strong emphasis on self-empowerment and efficient project management.

Peter Jaworowicz, PhD, MBA, PMP

Peter (*Piotr*) Jaworowicz is an accomplished program manager and marketing expert with over a decade of experience leading strategic initiatives for top tech and retail companies, including Google, Docusign, Salesforce, Walmart Global Tech, and Chevron. Throughout his career, he has successfully managed a wide range of marketing campaigns, research projects, and large-scale enterprise transformation programs.

Peter holds a PhD in Media Studies and an MBA in Management, with additional certifications in project management (PMP and PRINCE2).

He is also a certified Software Quality Assurance professional and completed Agile Product Owner training at Stanford University.

Peter is passionate about photography and contemporary art.

* * *

The Growth Mindset in Project Management is Magda's and Peter's sixth co-authored book. Prior to that, they collaborated on five books covering a range of marketing topics, including Copywriting, Political Marketing, Gender Marketing, Video Marketing, and Event Marketing.

They also have a history of project and marketing collaboration, having lectured together at universities and served as strategic consultants and training providers for executive teams across various industries, from retail to automotive.

* * *

Peter and Magda are married and reside in the San Francisco Bay Area, California, with their son, Nathaniel.

Unlock Your Full Potential with Growth Mindset Coaching

If you are eager to take your skills to the next level, there are several ways we can work together to deepen your understanding and practical application of a growth mindset in both professional and personal life.

Growth Mindset in Project Management: Live or Virtual Session

Whether you're leading teams, managing complex projects, or facing frequent challenges, this tailored session will help you:

- Build resilience in high-pressure environments
- Turn failures into learning opportunities
- Cultivate an adaptable and proactive mindset

Grow Your Mindset Workshops for Teams

This interactive workshop helps teams develop a growth mindset, reduce the fear of failure, and build a culture of continuous learning. With experience working with Fortune 500 companies, small businesses, and nonprofits, we tailor the workshop to suit any needs:

- **Creative workshops:** Perfect for smaller teams or startups, using games and activities to foster creativity and teamwork.

- **Corporate workshops:** Focused on strategic growth and team development, enhancing collaboration and performance.

Grow Your Mindset Coaching for Individuals

Exclusive one-on-one coaching program dives deep into understanding what drives you, highlighting your energy boosters and energy drains. We'll work on real-life examples from your work and personal life to support your goals and maintain work-life harmony. Benefits include:

- Deep self-reflection and goal alignment
- Techniques to overcome limiting beliefs and habits
- Practical toolkit for achieving balance and success

Inspirational Talk: Energize Your Team

Looking for a dose of motivation? This powerful talk blends the latest insights on mindset training, organizational transformation, project management, and self-development trends. It's designed to:

- Inspire action and creative thinking in tackling challenges
- Provide actionable strategies for cultivating a growth mindset
- Develop a culture of continuous improvement and innovation

Perfect for conferences, team meetings, or leadership events, this talk will leave your team energized and ready to embrace change.

Ready to Unlock Your Potential?

Let's take the next step together. By integrating a growth mindset, you'll empower yourself and your organization to achieve new heights.

To learn more visit: **www.magdajaworowicz.com**